Complete Solar Photovoltaics for Residential, Commercial, and Utility Systems

Copyright: Steven Magee 2011

Edition 1

Cover Picture: The Sun as reflected from a solar panel.

Complete Solar Photovoltaics © Steven Magee

Contents

1. Introduction — Page 4
2. Solar Cell — Page 8
3. The Basics of Solar Photovoltaics — Page 14
4. Photovoltaics and Weather — Page 19
5. Albedo — Page 35
6. Solar PV Systems — Page 41
7. System Selection — Page 47
8. Grid Interconnection Requirements — Page 80
9. Electrical Codes — Page 85
10. System Sizing — Page 88
11. Solar Modules — Page 100
12. Mono-Crystalline Silicon — Page 111
13. Poly-Crystalline Silicon — Page 118
14. Amorphous Silicon Thin Film — Page 125
15. Cadmium Telluride Thin Film — Page 132
16. CIGS Thin Film — Page 139
17. Solar Module Summary — Page 146
18. Inverters — Page 150
19. Switchgear — Page 153
20. Distribution — Page 154
21. Power Purchase Agreements — Page 157
22. Site Licensing Agreements — Page 158
23. Health and Safety — Page 159

Complete Solar Photovoltaics © Steven Magee

24. Planning Requirements	Page 179
25. Residential Design	Page 184
26. Commercial Design	Page 202
27. Utility Design	Page 228
28. Construction and Commissioning	Page 270
29. Tools Needed	Page 276
30. General Problems	Page 279
31. Solar Module O & M	Page 287
32. Mounting System O & M	Page 312
33. Cable O & M	Page 322
34. Combiner Box O & M	Page 324
35. Inverter O & M	Page 327
36. Switchgear O & M	Page 334
37. Distribution O & M	Page 335
38. Dirty Electricity	Page 340
39. Ancillary Systems O & M	Page 344
40. Summary	Page 354
41. References	Page 358
42. Electrical Symbols	Page 359
43. Electrical Formulae	Page 363
44. About the Author	Page 365
45. Author Contact Details	Page 372

Introduction

Complete Solar Photovoltaics is compiled from the first three books that I wrote on solar photovoltaics. It is a complete overview of the solar photovoltaic field that encompasses the training, design, and operations and maintenance aspects of the industry.

Solar photovoltaic technology has now become mainstream and is widely used around the world. All of the items needed to construct a reliable system can now be purchased from many different vendors and standards now exist in the industry to ensure that these different products can be seamlessly integrated together. The future of solar photovoltaics is bright and rapid adoption of the technology is underway.

Complete Solar Photovoltaics will teach the foundations of solar photovoltaics to those who are new to the field. For those already in the field, it should clear up any misconceptions that may have been formed regarding solar photovoltaics. Solar photovoltaics looks relatively easy to understand on the surface, but once you are involved with it on a large scale you will soon realize that there are many pitfalls to avoid. This book is aimed at making you successful and well informed so that you can be the very best in solar photovoltaics.

Solar photovoltaics is very different from conventional electrical system theory and because of this it has its own electrical codes that are contained in National Electric

Complete Solar Photovoltaics © Steven Magee

Code (NEC) Section 690 Solar Photovoltaics in the USA. We will explore the concepts of solar photovoltaics and you will become familiar with the differences between solar photovoltaics and conventional electrical theory.

Complete Solar Photovoltaics will demonstrate how to develop well built and reliable grid interconnected solar photovoltaic power systems across the different scales of the applications. As you will see, there really is not that much difference between a small residential system and a large utility scale system.

Complete Solar Photovoltaics should be read in conjunction with your local building codes, electrical codes, and grid interconnection utility codes. These vary widely and will be very much location dependent variables. Always consult with a Professional Engineer who is experienced in solar photovoltaics when building one of these systems.

Once designed and built, you will need to know how to operate and maintain grid interconnected solar power systems across the different scales of the applications. Solar photovoltaics has numerous problems that you should look for when operating and maintaining the systems. Many of these problems are unique to the field of solar photovoltaics and we will explore them in detail.

When maintaining and operating solar photovoltaic power systems, this book should be read in conjunction with your operation and maintenance manuals for your installed equipment. These vary widely and will be very much dependent on the manufacturer. Always operate and maintain your system in accordance with the manufacturers recommendations and follow the maintenance schedules in order to maintain warranty coverage. Work safe around the systems and use safety equipment as appropriate.

Solar photovoltaics has longevity with many companies offering twenty five year warranties on their solar modules. It is quite possible that you will only ever purchase one system during your lifetime and keeping it well maintained will assist with this.

The first editions of the training, design, and operations and maintenance books were written after I had just finished commissioning the DeSoto Solar installation in Arcadia, Florida. Opened by the President of the USA in 2009 as the largest solar photovoltaic installation in the USA, it was the thirteenth largest system in the world and the largest single axis tracking solar photovoltaic system in the world. Three hundred acres of solar power is an impressive sight!

Solar photovoltaic books that were written by engineers that had this type of experience were lacking and people in the USA solar photovoltaic industry were struggling to build effective and reliable grid connected solar power systems. The first editions were researched, written, and published relatively quickly to fill this gap in knowledge that was so apparent in 2009. Several other books also had to be produced in order to fill the gaps in knowledge that I was observing in the industry.

Complete Solar Photovoltaics © Steven Magee

Technology has been moving rapidly in solar photovoltaics and there are now several types of solar photovoltaic systems available. We will review the many systems types to enable you to work with them. By reading this book you will be on your way to building reliable grid connected solar power systems that work well from the very first day of operation.

This book is almost four hundred pages and it reflects the growing field of grid connected solar photovoltaics! I hope you enjoy it!

Work safe, follow the codes, and build great systems!

Solar Cell

The solar cell is the basic building block of the solar power system. On a system that may have 100,000 solar modules, you may have 7,200,000 solar cells! The typical solar module used today for grid connected solar power systems has 72 solar cells within it. You can see them when you look through the glass. Typically squares of crystalline silicon wafers of a few inches long.

The silicon solar cell was developed in the 1950's and it has been around for a long time. The equipment to convert the DC power that it produces was typically too expensive to produce large solar power plants until recently. That has now changed and we are seeing the solar cell be widely adopted as an effective producer of grid electricity.

The solar photoelectric effect was discovered by A. E. Becquerel in 1839. The properties of the solar photoelectric effect were explained by Albert Einstein in the 1905 while he was researching light. It is simply a semiconductor P-N junction that reacts to light. The more light you expose it to, then the more current that you will get out of it. The conversion of power is proportional to irradiance. Irradiance is the measurement of power of solar radiation in watts per meter squared (W/m^2).

The first solar module was built by Bell Laboratories in 1954. Solar power was widely adopted in the Space and satellite industry in the 1960's. Until the 1980's solar modules were simply too expensive to be widely adopted. Off grid homes and business adopted the technology in the 1980's. The 1990's saw the development of grid connected solar power systems that have become the prevalent application of the technology today.

The prices of solar modules have fallen dramatically and they can now be found for as little as $1.50 per watt. This price is half of what they were just a few years ago. World production of solar has ramped up and there are now many companies producing solar cells and solar modules. This will speed up adoption of the technology as the energy industry moves away from fossil fuels.

To design and maintain a solar power system, you need to understand that the basic building block of your system is the solar cell.

The following pages show the properties of solar cells.

The Solar Cell

This is the circuit symbol of the solar cell. You may also see it shown as a DC battery.:

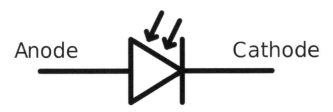

This is what a crystalline silicon solar cell looks like:

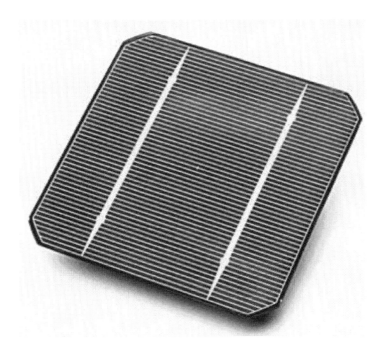

Solar Cell Temperature

The solar cell has a current limited output that proportionally matches the solar irradiance levels. When it is sunny it creates a lot of current, when it is cloudy it produces little current, and when it is dark it produces no current. The voltage that the solar cell produces is influenced by the temperature and it generally will produce 10% to 20% more voltage in wintertime as compared to summertime.

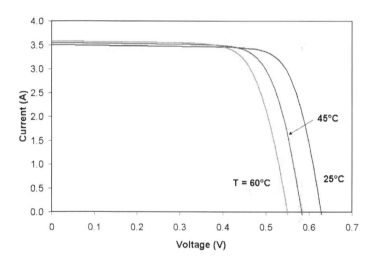

Solar Cell I-V Curve

As you can see, the solar cell is a current limited source that is dependent on solar irradiance. The more solar irradiance that there is, the more current the solar cell will produce.

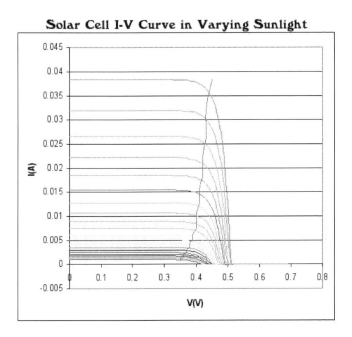

Complete Solar Photovoltaics © Steven Magee

MPPT Tracking

There is a peak area under the I-V curve that the maximum power from the solar cell is produced. This point is constantly changing with temperature and solar irradiance levels. As such, a process that is called maximum power point tracking (MPPT) is employed to get the maximum power out of the solar cell. MPPT automatically adjusts the current output from the solar cell so that the best combination of voltage and current is always obtained to ensure peak power from the solar cell.

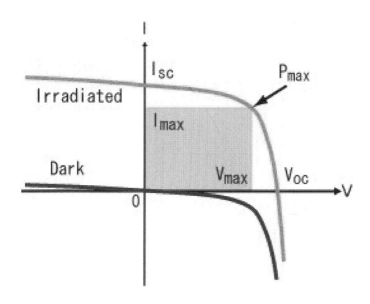

The Basics of Solar Photovoltaics

Solar photovoltaics comes in many different types:

- Monocrystalline
- Polycrystalline
- Thin film
- Other technologies

Although the technology is different between each type, they all do the same thing. All direct current (DC) solar modules generate DC electricity when exposed to sunlight. The listing above is in order of how efficiently each type will convert sunlight into electricity.

A typical direct current (DC) module will have the following electrical ratings on its label:

- Temperature adjustments
- DC Open circuit voltage (Voc)
- DC Maximum power point voltage (Vmpp)
- DC Short circuit current (Isc)
- DC Maximum power point current (Impp)
- DC Rated system voltage

Complete Solar Photovoltaics © Steven Ma

All of these values are given for Standard Test Con. (STC). Lets look at what each one of these mean:

Standard Test Conditions (STC)

Standard Test Conditions (STC) is how the solar module performs at a temperature of 25 °C, an irradiance of 1000 W/m² with an air mass 1.5 (AM1.5) spectrum. This is a standard test for all solar modules that are manufactured for the USA market that was developed by the photovoltaic industry and the government. It represents an average set of conditions that can be expected at the mid point between north and south of the contiguous forty-eight states during spring and fall with the Sun perpendicular to the solar module. San Francisco, California and Wichita, Kansas are near this midpoint of 37 degrees latitude. In Asia Seoul, Korea, is near and in Europe both Sevilla, Spain and Cantania, Italy are near to 37 degrees latitude.

It is important to note that a solar module output will be continuously variable during the year and even during the day. During the changing seasons it may output less power than its rating and sometimes it may output more power. These electrical ratings are for guidance only and it is where many new photovoltaic designers make mistakes in thinking that the module will never output more power than its rating. It is important that you understand that these solar photovoltaic modules can output far more power than their label states. It can be over fifty percent more and this will need to factored into the system design.

Temperature Adjustments

...dules are affected by temperature, both hot and ... adjustments to the module ratings needs to be ... the operating temperature outside of 25 degrees ... It is important when designing a system that the ...l temperature minimum and maximum values are for the area where the system is to be installed and ...djustments are factored into the design.

Open Circuit Voltage (Voc)

The open circuit voltage rating is how much voltage the module will put out with no load attached. This is an important value in order to design a system. This is the voltage to use when selecting your components and it must be adjusted for the historical minimum and maximum temperatures for the area. If more than one module is connected in series then multiply this temperature adjusted voltage by the number of modules in series to get the total maximum DC voltage of the system.

DC Maximum Power Point

The DC maximum power point is a simple concept. Power is a function of both voltage and current. The maximum power point is obtained when the current and voltage from the module when multiplied together give the maximum power figure. These values will change constantly during the day with the weather conditions. Voltage will remain relatively constant, but current will vary a lot with irradiance. The DC to AC inverter system constantly monitors the power from the solar photovoltaic DC system and automatically keeps the inverter system working at the maximum power value for the given conditions.

DC Maximum Power Point Voltage (Vmpp)

The DC maximum power point voltage (Vmpp) is the operating voltage of the solar module under load. Again this value will change with temperature and irradiance, but should only vary by about twenty percent of the STC rating during the day time.

DC Short Circuit Current (Isc)

The DC short circuit current value is the maximum current that the module will output at standard test conditions if the positive and negative terminals were connected (shorted) together. It is important to note that this value will vary a lot dependent on weather conditions and can be over fifty percent larger.

DC Maximum Power Point Current (Impp)

The DC maximum power point current is the amperage that the solar module will output at standard test conditions in normal operation. It is important to note that this value will vary a lot dependent on weather conditions and can be over fifty percent larger.

DC Rated System Voltage

This is a very important design value. It is a rating of how many modules can be safely connected together in series, this is called a solar photovoltaic module string. This system voltage should never be exceeded when adjusting for the minimum and maximum temperatures of the area that the system is being installed. This value basically limits the number of modules that can be connected in series in the system.

Photovoltaics and Weather

The performance of any solar photovoltaic system is dependent on the weather. The main factors that affect the system performance are clouds, irradiance, temperature, shade, latitude and how dirty the solar modules are. Let's now explore the effects of the weather in more detail:

Irradiance

Irradiance is a measure of how much sunlight the solar module is receiving. It is given in watts per meter squared or W/m^2. Standard Test Conditions (STC) uses a value of $1,000W/m^2$. This value can range from $0W/m^2$ at night through to over $1,500W/m^2$ during a day interspersed with large fluffy clouds. This value of $1,500W/m^2$ is larger than what you would receive in space. The reason why we can get greater values at ground level is due to what is known as the "cloud effect". Normally the sunlight is traveling in a straight line from the Sun to our solar module with some atmospheric scattering. However, when clouds are present they can also reflect and can act like lenses to send some extra sunlight onto the solar modules. This extra light is converted into extra energy and this is seen largely as an increase in power from the system. This effect can be a few minutes long in duration when it occurs. The diagrams on the following pages demonstrate the "cloud effect".

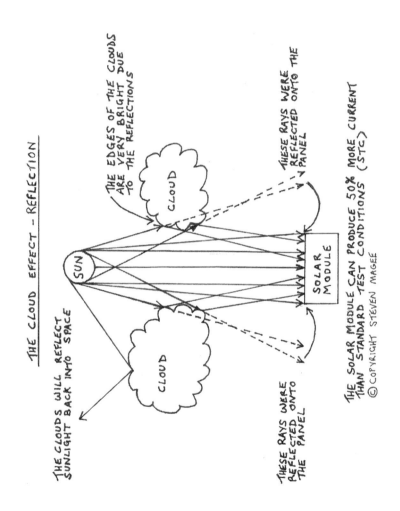

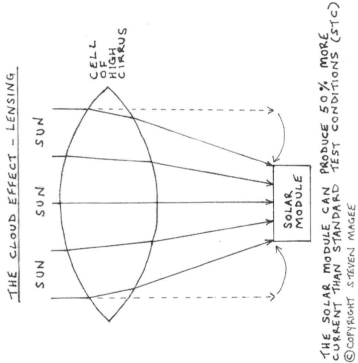

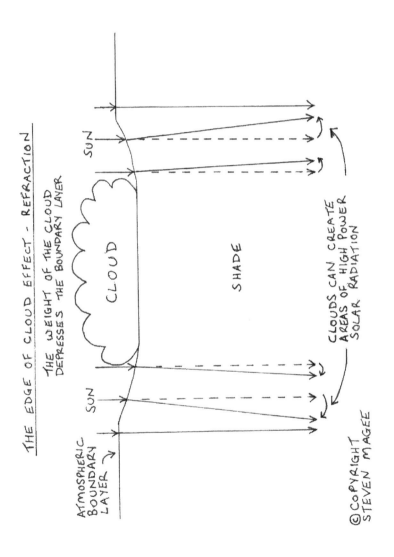

Complete Solar Photovoltaics © Steven Magee

The various kinds of emissions and smoke can affect solar power radiation transmission to the ground. Smoke causes atmospheric turbulence to occur and this can create lensing effects. You should be wary of installing solar power systems in areas that have smoke and various atmospheric emissions nearby as you may find that it starts to cause problems with your solar photovoltaic system.

Heat haze causes a similar effect to smoke and emissions. Heat causes turbulence to occur and this may cause lensing effects. You may see surges in power on a system if there is a cell of heat above it.

Both of these effects may occur in cities with tall buildings. The buildings cause turbulence to occur and may create lensing effects that cause the power to surge on photovoltaic systems.

These effects are shown in the following diagrams.

Complete Solar Photovoltaics © Steven Magee

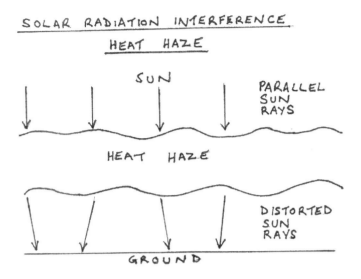

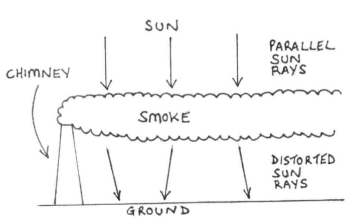

© Copyright Steven Magee

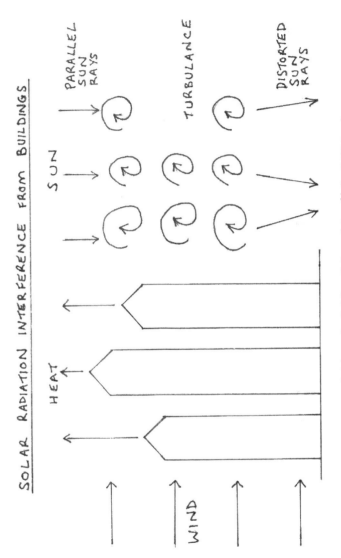

Other effects on irradiance are the snow effect, the water effect (lake/ocean/wet surfaces after rain), the building effect and albedo. Snow cover, water, glass covered buildings, reflective painted buildings and roofs, and the albedo of the area surrounding the solar photovoltaic modules can reflect extra sunlight onto the solar power system. If you are installing a system in an area that has any of these, it is important to account for it. Each effect can produce an increase in power output. If you find yourself having to wear sunglasses in the solar photovoltaic system location for your eyes to be comfortable, then you probably have light reflections taking place.

During the seasons the system may operate at below the standard test conditions values and at other times it may exceed these values. During the design phase of the system you will need to assess where the greatest power need is and perhaps increase the size of the system accordingly for the particular time of year.

Air Mass

Air mass is a measurement of the amount of atmosphere that the sunlight has to pass through to get to the ground. It varies with the seasons and also the location on the Earth. Within the tropics, air mass will reach its maximum power value of 1 during summertime. Air mass 1 corresponds to the Sun being directly overhead, air mass increases as the Sun moves from directly overhead down to the horizon.

All USA solar modules are rated for air mass 1.5 which corresponds to a central USA location in spring and fall. When in a southerly location you will approach air mass 1 which will increase power output by about 13% from STC in the USA.

Locations that are at or near air mass 1 in summertime in the USA are all Hawaiian islands, Florida and Texas. Approximately half of the continental USA is located between air mass 1 and air mass 1.5 in spring and fall. If you are working on systems that are located in these southern USA states, you will get more power out of these systems due to a decreased air mass. In summertime the air mass will move closer to 1 in the continental USA.

Clouds

Clouds come in many forms. An important question is how do clouds affect irradiance on solar power photovoltaic power systems? The list below will help with understanding the effects of clouds on irradiance at air mass 1 (within the tropics in summer time):

- Clear, sunny skies will give approximately 1,130W/m^2. The transmission characteristics of the atmosphere will vary in clear skies, sometimes being relatively transparent and other times being more opaque and this affects irradiance values. Air quality is a major factor for the transmission of sunlight through the atmosphere. Particulate matter in the atmosphere will reduce the transmission level.

- Thin cirrus will give approximately $1,000W/m^2$. Thin cirrus will give even and relatively stable irradiance levels due to scattering of the light.
- Thick cirrus will give approximately $750W/m^2$.
- Thin clouds will give about $500W/m^2$.
- Thick clouds will give about $250W/m^2$. No shadows on the ground will be present
- Thick clouds with a visibly dark sky will give about $100W/m^2$. No shadows on the ground will be present. You will not be able to see the location of the Sun in the sky.
- Broken clouds will give surges of about $1,500W/m^2$ and reductions to about $100W/m^2$ of irradiance due to the cloud effect. The rate and length of time for these surges and reductions is dependent on the speed of the clouds passing in front of the Sun.

Temperature

Temperature will affect the system to a much lesser extent than irradiance. The cooler the system is below 25 degrees Celsius, the more power it will produce. Correspondingly, the hotter the system is above 25 degrees Celsius, the less power it will produce. Temperature can affect solar photovoltaic system power output by about twenty percent. Temperature will cause expansion and contraction effects to occur and these should be designed into your system. Aluminum rails can expand 1" over 60' with a 100 °F temperature change!

Shade

It is undesirable to shade solar photovoltaic modules as it can significantly affect the performance of the system. When studying the location of where to install a system, always factor in the surroundings for shading effects. Avoid shading with solar photovoltaic power systems.

Wind

Wind will provide cooling to the photovoltaic modules and it is an aid to power production. A breezy location will provide improved performance from the system. When mounting solar modules onto racking, it is good to allow spaces between the solar modules in order to aid with cooling airflow around the modules and also to reduce wind resistance. When choosing solar modules and mounting systems, it is important to ensure that they are rated for the wind speed of the area that you are installing them into. This can be obtained from the Universal Building Code (UBC), International Building Code (IBC) or from the American Society of Civil Engineers (ASCE) standard 7-10.

Altitude

A higher altitude location will improve the amount of irradiance that the system will receive, due to less scattering and absorption of the sunlight by the atmosphere. It also acts as a natural cooler of the system which further improves system performance. Generally a high altitude location will have a higher percentage of clearer skies during a year which will give a higher energy yield from the system. High altitude sites will suffer from accelerated aging on the equipment. Many companies will not honor the warranty above certain altitudes due to this.

Snow and Ice

Snow and ice may affect a solar module if it is faulty, causing the glass to break. It may obscure its view of the Sun. Tracking systems can be affected by this and in some snowy locations it is advisable to park the tracking system facing South during these periods. The reflection from the snow will increase the power from the system in wintertime. Snow loads can be obtained from the American Society of Civil Engineers (ASCE) standard 7-10

Hail

Hail can break solar modules, so it is important to know type of hail that your area can receive. If you get large golf ball size hail, you may not want to install glass solar modules. Solar modules are tested for hail and pass the tests even if the glass module breaks. The test just ensures that the modules remain intact when broken. Glass solar modules are hard to break and normal sized hail should have no effect.

Dirt

Clean solar modules are the desirable configuration for a system. However, dust and dirt will get onto the surface of the modules and will degrade performance by up to 10% on average. Cleaning the modules is very much a function of the location where they are installed and also how dirty they are. Most people will clean on an as needed basis, generally when they are visually very dirty. Always follow the manufacturers instructions for cleaning your particular modules and remember that solar modules are operating with electricity flowing in them when exposed to light. Night time cleaning is recommended for safety.

Lightning

Lightning can affect solar modules, especially on large systems that cover fields. Good equipment grounding is the way to deal with this threat. A low resistance ground will generally dissipate lightning away from a solar module that is struck by lightning. Generally, the damage should be limited to only the solar module that was struck. If a cable is struck, then lightning surge arrestors can limit the damage in the system. These are generally installed in the inverter and on larger systems, in combiner and re-combiner boxes. Lightning may blow the string fuse/circuit breaker(s) for the module(s) struck. Install lightning protection as recommended by the manufacturers of the products used in the installation.

Seasons

We have four distinct seasons of winter, spring, summer and fall. We can word this another way as winter solstice (December 21), spring equinox (March 20), summer solstice (June 21) and autumn equinox (September 22). What does this mean to a solar power system?

- The length of the day
- The angle of the Sun (air mass)
- Heating and cooling
- Rain
- Albedo

Winter solstice is the shortest day of the year and summer solstice is the longest day of the year. Spring and fall equinoxes are when day time is the same length of time as night time.

Regarding the angle of the Sun in the USA, winter solstice is when the Sun is at the lowest in the sky, or 23.5 degrees below the equator and summer solstice is when it is 23.5 degrees above the equator. Spring and fall equinoxes are when the Sun is directly overhead at solar noon at the equator.

For our solar power system, this means that we will produce our largest voltage in wintertime when it is the coldest and we will produce our largest current when it is peak irradiance and albedo combined.

Complete Solar Photovoltaics © Steven Magee

The changing seasons will affect rainfall and in dry seasons you may want to schedule cleaning to keep the modules in good performance. Rain generally helps to keep the module clean naturally. Rain will also cause the albedo to change around the solar photovoltaic system and you will need to take this into account.

The albedo of the site will change during the seasons and you will need to factor this into your design. A barren snow covered field will be a lot different to one filled with corn or flowers.

There are a number of things to consider with the seasons:

- Spring & Autumn
    - The system will be operating close to standard test conditions (STC) and measured values should be close to that on the solar module label.
    - This is the most favorable time for outdoor working.
- Summertime
    - The system will be hot and the DC voltage will be lower than normal.
    - Ambient temperatures will be high.
    - Heat and dehydration may be a problem for working on the system.
- Wintertime

- The system will be cold and the DC voltage will be higher than normal.
- It may be too cold to work on the system.
- Frost, ice and snow may be an issue for performing maintenance.
- Ambient temperatures will be low.

Due Diligence

It is important when designing, operating and maintaining a solar power generation system that you are aware of the annual climatic conditions to expect. Amongst the data that you should have is:

- Historic annual minimum temperature
- Historic annual maximum temperature
- Historic annual maximum wind speed
- Historic annual snow fall depth
- Historic annual hail size
- Historic annual peak irradiance
- Historic monthly irradiance
- Historic annual peak albedo
- Historic monthly albedo

With these values you will be able to make educated engineering decisions regarding the selection of your system.

Albedo

Albedo is the Siamese twin of the cloud effect. They go everywhere together. Make sure that you are considering both effects when estimating solar radiation power levels. Albedo is the reflectivity of surfaces. A very well known effect in the world of astronomy.

Everything reflects light, even matt black surfaces reflect some low level of light, that's how we can see it. Everything our eyes can see is created from reflected light from surfaces. If your eyes can see these things, so can the rest of your body. If you need to put on your sunglasses, think about the reflection effects that may cause your eyes to be uncomfortable. Your body will sense the increase of solar radiation power levels as heat and you may start to sweat as your body warms up. You will need to identify these effects as it may be able to impact your solar power system.

You should get used to assessing your environment for these albedo effects and start to avoid them if possible. Nature suppresses solar radiation for a reason. These albedo effects will probably put your body into an environment that it was never designed to cope with and the long term effects may be undesirable. The short term effect will be sunburn and possibly heatstroke.

Complete Solar Photovoltaics © Steven Magee

Here is a list from Wikipedia that shows the albedo in various objects:

Object	Albedo
Fresh asphalt	0.04
Worn asphalt	0.12
Conifer forest	0.08 to 0.15
Deciduous trees	0.15 to 0.18
Bare soil	0.17
Green grass	0.25
Desert sand	0.4
New concrete	0.55
Ocean Ice	0.5–0.7
Fresh snow	0.80–0.90

Complete Solar Photovoltaics © Steven Magee

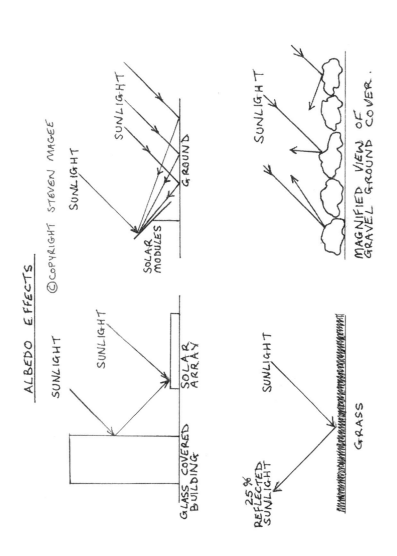

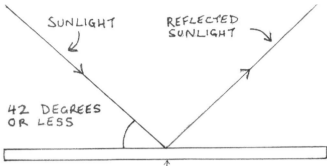

SOLAR MODULES BECOME VERY REFLECTIVE AT 42 DEGREES OR LESS. THEY WILL PRODUCE LOWER POWER AT SUNLIGHT ANGLES OF LESS THAN 42 DEGREES. YOU SHOULD BE AWARE OF THIS WHEN DESIGNING, OR TROUBLESHOOTING POOR PERFORMING SYSTEMS.

© COPYRIGHT STEVEN MAGEE

Complete Solar Photovoltaics © Steven Magee

The albedo effects will be at their strongest around:

- Glass covered buildings.
- Light painted buildings and structures.
- Summer solstice.
- Clear atmospherics.
- Snow.
- Water.
- High altitude.
- Cloud effects.
- Tilted modules that can see the ground.

Each of these can increase the short circuit current past the labeled value on the solar module. In some cases, it may be sufficient to blow the solar module (string) fuse. You should be expert in predicting albedo before attempting to design a solar power system.

The next page shows one of the most problematic albedo causes, the mirrored building!

Complete Solar Photovoltaics © Steven Magee

The "Multiple-Sun" Effect

Building glass can create the "Multiple-Sun" effect.

Solar PV Systems

Grounded Systems

Grounded systems are the majority of USA installations. Basically, one conductor of the array is always grounded. There are no fuses or disconnects that can interrupt this grounded conductor. DC solar fields can either be positive grounded or negative grounded. The grounded conductor is white and the ungrounded conductor is black.

Ungrounded Systems

Ungrounded systems have both the positive and negative conductors fused and switched. Their design and construction is almost the same as conventional grounded systems except that the inverter does not ground any of the current carrying conductors. Instead it leaves the array floating and this eliminates the need for an internal inverter transformer and improves system efficiency. The two current carrying conductors should be marked red for positive and black for negative. The equipment grounding is identical to a grounded solar photovoltaic system. It costs more to construct these systems due to the extra fusing and DC disconnect switches that break both conductors that are required.

Bi-Polar Systems

Bi-polar systems have a positive grounded solar field and a corresponding negative grounded solar field of equal size. This enables the inverter to not have an internal transformer and increases the inverter efficiency. The grounding takes place internally in the inverter.

A bi-polar system diagram is shown on the next page.

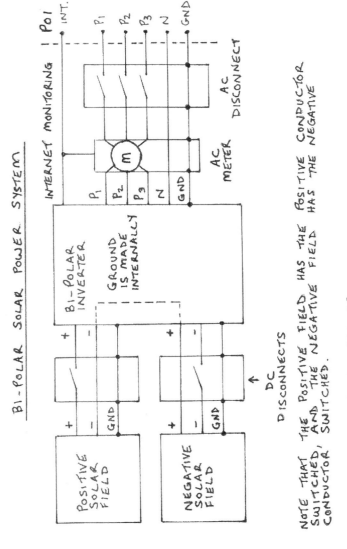

Split Systems

Split systems convert the variable voltage DC that comes from the modules into a set voltage DC that is near the 600V/1000V limit for DC systems. It enables the central DC to AC inverter system to be simpler and more efficient. By having a high voltage DC system, it allows cable sizes to be smaller and improves efficiency. DC to DC conversion can take place at the module, string or combiner box level. These systems facilitate the addition of battery storage at a later date.

String Inverters

String inverters can improve the efficiency of the system and should be considered where strings of solar modules may be pointing in different directions, or get different levels of shading during the day.

Module Inverters

Module DC to AC inverters are available and should be considered in areas prone to shading or residential systems where there is no space to mount a larger inverter system. The micro-inverters are mounted to the racking system, underneath the modules. They plug into each other with factory assembled cords. You should check with the manufacturer of the micro-inverter for a list of compatible solar modules. A system using micro-inverters can be built over time and as money permits, as there are no string sizing requirements. Module mismatch losses are eliminated with micro-inverters.

AC Solar Modules

AC modules are becoming available where the micro-inverter is actually part of the solar module. These simplify the installation with no DC circuit design needed. It is a factory assembled solar power system and it really does not get any simpler than this. Just parallel the AC solar modules to increase the system power. Generally about 15 modules can be connected in parallel on each AC circuit.

AC solar modules and micro-inverters are shown on the next page.

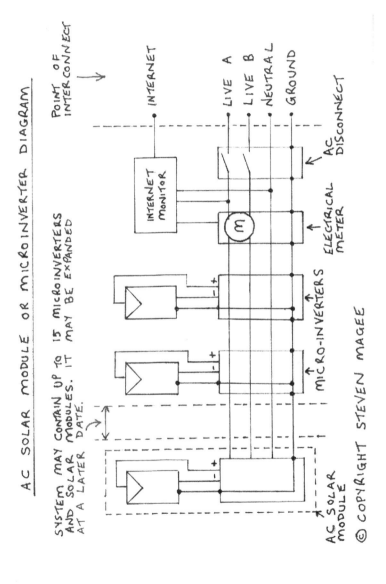

System Selection

System selection is the most complicated stage of solar photovoltaic system designs and will determine how well your system works. When done well, your system will give years of trouble free operation.

Solar Modules

Solar modules typically represent the largest investment in the system. Traditionally silicon solar modules have been used, but now many new types of technologies are emerging such as thin film, and so on. All solar modules are tested to the Underwriters Laboratory (UL) 1703 standard in the USA.

Crystalline silicon is the best understood, the most efficient and has been around for many decades. The newer film type solar modules are less efficient and cheaper to purchase. Unfortunately more thin film modules are needed to generate the same power as crystalline silicon, and this increases the system physical size and associated support systems such as cabling, racking, installation costs, and so on. Thin film is a relatively new technology and, as such, presents a higher long term risk for systems that are designed for at least 25 years of life. Temperature affects film systems much less than crystalline silicon and it is about a factor of two difference.

Complete Solar Photovoltaics © Steven Magee

Generally any decision on which technology to use is driven by market rates for each type of technology, aesthetics and personal preference. Solar modules are a commodity and their prices can fluctuate rapidly.

Solar modules should have a temperature adjustment applied to them that indicates their actual operation temperature. A solar module in operation in summertime can be 40 °C hotter than ambient. Solar module calculations should be adjusted by the following values for accurate calculations:

+40 °C for a roof mounted system

+30 °C for a tracking system or ground mounted system

Frame-less solar modules are available and these generally offer better performance in shedding snow, dust and dirt. However they generally have lower wind speed ratings and are more prone to breakage during system construction and physical damage once in operation.

Separate inverter systems should be used for modules that face different directions or ones that may get shaded during the daytime.

Verify if the modules can be used in both positive and negative grounded systems. Some modules can only be used in either positive or negative grounded systems. They will not work in both configurations. High efficiency modules are where you see this problem in general.

Module tolerances are typically +/- 10% for both current and voltage values. Power can be +/- 5%. Pay attention to the power tolerance values. There are many manufacturers that have wide tolerances and you may not get the power from the system that you expect if they are all supplied as the lower tolerance. Try to use modules that have no negative power tolerance to avoid this problem. Module mismatch losses come from these differences in values.

You should consider using all black solar modules on residential systems that can be seen from the ground, as it will improve the aesthetics of the system. All black modules will run a little hotter, but will look better from the ground. You should mount the modules in the same plane as the sloped roof, as pitched modules do not have curb appeal.

Building integrated photovoltaics (BIPV) are a bad idea if it puts the solar modules into the human environment. Solar modules produce electromagnetic interference (EMI) in operation and this may make the people in the building sick. Solar modules should be kept out of the human environment and should be in areas where people are not routinely present. EMI passes through most building materials easily and distance is the only effective way of reducing exposure to it.

When sizing strings of solar modules, if you have leeway with the choice of string size, choose the largest number of modules for the string. This ensures that as the system ages that you will not fall below the inverter low voltage DC threshold.

Solar modules can be mounted in any orientation. However, it is preferable to mount solar modules with the cables coming out of the bottom of the module junction box. This will prevent rain and condensation from draining along the cables towards the junction box. This would normally be a portrait orientation.

Solar modules decrease their output over time and can be producing only 80% of their labeled value after 25 years. If power production predictions are important to you, decrease the DC power by 1% per year to account for this module degradation.

Mounting System

There are three ways to mount your modules:

- Fixed tilt
- Single axis tracker
- Dual axis tracker

All have their pros and cons.

The fixed tilt system is the most common and is widespread. The solar modules are either mounted to a roof, building or are ground mounted in a fixed position inclined to face south at a tilt angle matching the latitude.

Some systems allow you to adjust the tilt angle of the modules for the season, but it appears that most people prefer the low maintenance option of mounting the modules into a fixed position for the entire year. The fixed tilt system is the most reliable configuration and also the lowest cost. Washing costs increase the closer to horizontal that the module is mounted as dirt will build up quicker. The downside is that it has the lowest annual energy output of the mounting systems.

A simple fixed tilt system is shown on the next page.

Fixed Tilt

A simple fixed tilt solar photovoltaic power system.

Complete Solar Photovoltaics © Steven Magee

The single axis tracker works well in. The solar modules are mounted on a rotating north-south axis which allows them to track from east to west during the day. There are two types of single axis trackers generally available. The first has the north-south axis mounted horizontal and the modules can track in the east to west direction. This system works well in or near to the tropics where the Sun can be almost directly overhead. The second has the north-south axis inclined to match the latitude and this enables the solar modules to face the Sun in spring and fall. This system works better as you move further away from the tropics. A single axis tracker can increase power output by about 20% when compared to a fixed tilt system. The single axis tracker does not cost much more than a fixed tilt system and the extra expense is generally offset by the extra annual energy yield of the system.

A car park solar canopy is shown in the next picture that uses single axis tracking to follow the Sun.

Complete Solar Photovoltaics © Steven Magee

Single Axis Tracker

The car park solar canopy uses single axis tracking to follow the Sun.

The dual axis tracker has the modules tracking the Sun from sunrise to sunset, keeping the solar modules in the optimal position for maximum power generation. A dual axis tracker can increase annual energy output by about 30% when compared to a fixed tilt system. The downside to a dual axis tracker is that it requires a lot of space, can be very tall, has a complicated control system, they are expensive, and they are the highest maintenance system.

The foundation systems can vary between the various systems and the soil analysis in conjunction with the wind and snow loading will generally dictate the foundation required. Common foundations are poured footings, driven steel piles or I-beams, helical piers, and ballasted. Ballasted mounts can typically only be used on grades of 5% or less.

All cuts that are made in galvanized steel mounting systems should be treated. In some areas you should consider using theft prevention mounting hardware.

For roof mounted systems, you should inspect the roof and if it is due for replacement, it should be done before the solar power system is mounted to it. If the roof is replaced, then work with the contractor to ensure that the mounts and roof penetrations are installed at the same time. If the roof is good, ensure that you are not voiding any existing roof warranties that may be invalidated by the mounting of a solar power system to it. Stagger the roof mountings so that they load the rafters evenly.

Complete Solar Photovoltaics © Steven Magee

Keep a good 3 feet of space around the edges of the solar power system for safe roof access. Mount the system as high up on the roof as possible to keep the effects of electromagnetic interference (EMI) out of the human environment. You should install permanent roof anchors for fall protection for both construction and maintenance purposes. Remember that the modules need to be cleaned and maintained and you should consider putting walkways in on large systems to effect this. You cannot walk on solar modules as they will break!

Be wary of the material that you choose to use for ground cover under your racking system, as these may impact your solar power system. Small pebbles and sand around solar power systems will get picked up by the wind and may scratch the surface of the solar modules. This may weaken the module and may increase the possibility of the glass breaking. Light colored ground materials may cause the power output to increase from the system due to albedo effects.

Grass can be used under mounting systems but it does need constant maintenance. Sheep have been used in solar power systems to keep the grass short, but do not use goats as they will climb on the equipment and eat the cabling! Mowing the grass and weeds may cause rocks to be thrown at the modules and you should expect breakages.

The diagram on the next page shows the differences for each tracker system at noon with the seasons.

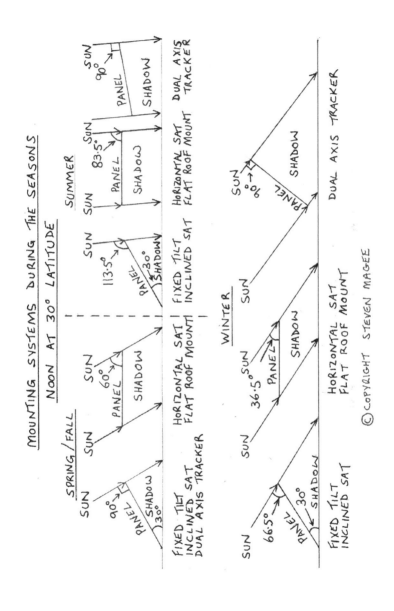

Cabling

For all outdoor cabling exposed to the Sun, solar rated cable should be used due to its resistance to ultraviolet radiation. UL 4703 covers the requirements for solar outdoor cabling. Once inside a building, conduit or underground this requirement need not apply.

On ground mounted installations it is important to remember that animals will be able to access the cables and equipment, so you will need be familiar with your local wildlife in the area. Where possible, enclose all accessible cabling with protective covers.

When you have excess cable, do not coil it up! You will make an inductor if you do this. Instead, just neatly place it into a U shape. You should also be running both the positive and negative DC conductors together to cancel out the fields that they generate. Ideally, the positive and negative DC conductors should be twisted together.

Conduit or ducting is recommended for all underground runs as it can be easily rewired to a larger size if you find that you system is generating more power than designed for or if a cable goes faulty.

The DC cable voltage drop should be within 3%. The AC cabling voltage drop should be within 1.5% . The AC cabling voltage drop is tighter due to the minimum and maximum voltage regulation limits that the inverter can operate in without shutting down. If you have a large voltage drop on the AC cabling, then you may see the inverter hitting its limits during high solar irradiance periods. The inverters generally have to be within 12% below and 10% above of their grid AC voltage rating or they will shutdown. On a 240v AC system, this equates to 211 volts to 264 volts.

NEC chapter 9 table 8 details DC cable resistance and table 9 details AC resistance and reactance.

When building a system with long cable runs, you should use the following voltages in the order presented for efficiency:

- High voltage
- Medium voltage
- 1,000V DC
- 480V AC 3 phase
- 600V DC
- 240V AC 3 phase or 277V AC 1 Phase
- 208V AC 3 phase or 240V AC 1 Phase

All cables should be able to carry 125% of the continuous load and 100% of the non-continuous load per NEC 210.19(A)(1) and NEC 215.2(A). The continuous load is defined as over three hours in duration.

The diagram on the next page shows how the string wiring should be run.

SOLAR MODULE STRING WIRING

CORRECT

PANEL OF 12 SOLAR MODULES

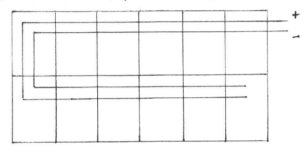

POSITIVE AND NEGATIVE WIRING SHOULD BE RUN TOGETHER TO REDUCE EMI EFFECTS. IT IS PREFERABLE TO TWIST THEM TOGETHER.

INCORRECT

PANEL OF 12 SOLAR MODULES

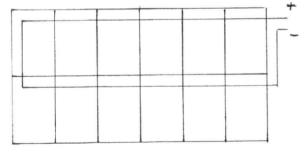

CREATING LARGE LOOPS CAUSES EMI EFFECTS TO BE WORSE. DURING LIGHTNING STORMS LARGE CURRENTS MAY BE INDUCED INTO THE WIRING, CAUSING FUSE AND MODULE DAMAGE.

© COPYRIGHT STEVEN MAGEE

Inverters

Inverters come in many different types and sizes. This book is dedicated to grid tie inverters and this is what we will consider. All residential and commercial solar photovoltaic grid tie inverters have to comply with UL1741 in the USA. This ensures that all grid tie inverters meet the requirements of the utility authority. Some of these requirements are below:

- Disconnect from grid during power cuts.
- Protection against faults.
- Good power quality output.
- Disconnects from the grid if there are power quality issues.
- Ground fault detection.

Utility inverters are distinctly different and we will discuss the differences in the utility design chapter.

When selecting an inverter system, you should be aware that by running it continuously at full power may lead to it running hot and power limiting and this may lead to a shorter life. You should be sizing inverters to convert 100% of the annual peak continuous current from the DC system for longevity. If you run them at more than 100% loaded then you will be throwing power away, you will not be MPPT tracking, the equipment will be under the maximum stress, and it may shutdown for the day if you exceed the maximum voltage. Overloading an inverter is a bad idea, and unfortunately, many people do it.

Complete Solar Photovoltaics © Steven Magee

Exceeding the maximum voltage may invalidate your warranty, as the inverters record the maximum voltage in permanent memory that can be read by the manufacturer.

Many people recommend a DC power system that is between 80% to 120% of the AC power rating. You should not exceed the 120% value. It should also be derated for altitude, as the inverter cooling will not work as well. At altitude, you may want the upper limit on the DC power to match the AC power. The more you load an inverter with DC power, the more problems you invite into your design and the life of the system.

Indoor inverters should be in air conditioned environments and large outdoor inverters should be shaded for maintenance purposes, rain, and sunlight protection.

Grid connected inverters commonly come in the following voltages:

- 240V AC single phase. (Residential)
- 208V AC three phase.
- 240V AC three phase.
- 480V AC three phase.

The voltage drop on the inverter AC cables should not be large enough under full load for it to hit the high voltage limit. Consider increasing the size of AC cables to reduce the voltage drop.

Transformers

Residential and commercial systems commonly have the transformer built into them during manufacture. This enables the inverter to be connected directly into the electrical system. An exception to this is transformer-less inverters that use bi-polar and isolated DC solar fields. These can connect directly into the electrical system without a transformer.

On utility systems you will most likely be installing transformers to enable your solar power to enter the utility grid system. There are many types of transformers available and the type to use can be confusing. Here are the main types that you will be considering:

- Pole mounted.
- Ground mounted.
- Dry.
- Oil filled.
- Transformer winding configuration.

Dry transformers sound appealing, but they can have problems with cooling, particularly if there are harmonics on the system. Oil filled transformers are usual choice.

The wye-delta, or star-delta in Europe, is the most common configuration of transformer. The delta winding in particular has the ability to filter the third and fifth harmonics that the inverters may generate. You should consult with the utility to establish the winding configuration that they prefer.

Transformers are available that have been constructed to act as a harmonic filter and they are a good match for the alternate energy industry. The alternate energy industry is basically using high powered switching electronics to drive a tuned filter to produce AC electricity. It is next to impossible to eliminate harmonics from these systems. You should evaluate these products as part of your transformer selection process, as it may assist you in meeting the power quality requirements of the electrical grid.

Generally for large utility projects you will use transformers that have two separate windings on the low voltage side for connecting in two inverter systems. This enables inverters to be connected in that do not use internal transformers and increases system efficiency.

If the inverter system has a built in transformer, then you can use a larger transformer and connect multiple inverters into it through the appropriate switchgear.

For electronic power conversion systems (PCS), commonly called inverters, a electrostatic shield should be included in the manufacture of the transformer to reduce the harmonic content.

You should pay attention to the transformer efficiency at various loads and also the no load losses. Try to select a high efficiency transformer that has low no-load losses.

The transformer will spend two thirds of its life operating in the no-load configuration during night time and low irradiance levels near sunrise and sunset. These no-load losses can add up. Typical transformer efficiency should be above 99%.

During fault conditions the transformer impedance will limit the fault current. Most utility transformers have an impedance of about 5 ohms. You should verify the transformer impedance during your equipment selection process.

The following pictures show the transformer types.

Complete Solar Photovoltaics © Steven Magee

Pole Mounted Transformers

Pole mounted transformers can save space and are generally cheaper to repair.

Complete Solar Photovoltaics © Steven Magee

Ground Mounted Transformers

Ground mounted transformers are generally easier to service. You will need to build in a trough to the foundation that can catch the oil if it were to leak.

Complete Solar Photovoltaics © Steven Magee

Balance of System (BOS)

The balance of system are the other items that are required to complete the installation. It is the mounting hardware, cable management systems, cables, combiner boxes, and so on.

The connectors that are typically used in the DC circuit are commonly the MC4 type or equivalent. There are many different connector types available and you should ensure that your modules are fitted with the same connectors that your strings cables and combiner boxes are using.

If you can, use DC combiner boxes that have a DC disconnect built into them. This simplifies maintenance in the field.

Using fuses or circuit breakers that are either too small for the application, under-rated, in hot locations, in fuse holders that are under-rated, can all lead to nuisance tripping. An over-temperature fuse or breaker may operate at a lower level than its labeled rating. Using them in circuits that exceed their interrupt ratings may lead to fires, exploding fuses, and personnel injury.

Account for expansion and contraction in all systems used. For conduits, use expansion couplings. All outdoor mounted equipment, terminals, and cables must have at least 90 °C ratings.

When sizing current carrying conductors, you must size for the larger of 125% of continuous current and 100% of non-continuous currents. Apply the appropriate temperature correction factor and conduit fill factor.

Wiring losses should be less than 3% and across the entire system you should aim to stay below 5% wiring losses. You should be aware that a solar power inverter system is rarely running at 100% and that power losses from wiring will be lower for most of the year.

System Costs

The typical system costs today are:

40% Solar modules.

16% Inverters.

12% Mounting system.

9% Taxes and fees.

17% Labor.

6% Balance of system costs.

Solar module prices are at record lows and may even continue to reduce in price, so the system costs are constantly changing. Copper and aluminum prices fluctuate rapidly and you should always have the current pricing for these items.

System downtime losses can add up and for this reason you will want your design to work well from day one. A bad design will cost you a lot of money in lost energy production and repairs over the lifetime of the system.

Solar Design Hints and Tips

All DC Grid Connected Solar Photovoltaic Systems Need

- Equipment Ground (European earth connection).
- DC Disconnect.
- Inverter.
- AC Disconnect.

As They Get Larger They Also Need

- Combiner box.
- Recombiner boxes.
- Transformers and switchgear.
- Distribution and transmission systems.

Important Guidelines for Tracking Systems

- Try to put no more than one tracking system on one inverter.
- Consider installing blocking diodes in each string to prevent reverse current flowing on multiple tracking systems installed on one inverter.

Roofs

- Do not mount solar modules onto flammable roof surfaces due to the high temperature arcing that may occur during fault conditions.
- Use south facing roofs where possible.
- When mounting solar modules to multiple roof surfaces that face different directions, you should be using micro-inverters.
- If the roof has shading during the peak solar hours you should be using micro-inverters.
- Use a mounting system that raises the solar modules at least 8" above the roof surface for efficient cooling.
- Familiarize yourself with the wide range of roof mounting products.
- Use snow guards on the bottom edge of the array to prevent large sheets of snow from sliding onto people below.
- Ground the mounting system prior to installing the solar modules.

DC component selection

- Always make sure that DC rated components are used.
- Components must exceed specifications for maximum voltage.
- Components must exceed specifications for maximum continuous current.

- All outdoor cables must be solar rated.
- All outdoor connectors must be solar rated.
- De-rate equipment, cables and fuses for installed conditions such as:
  - High ambient enclosure temperatures.
  - Burial.
  - Enclosed conduits and ducts.
  - Heating effects.
  - Cooling effects.
  - High power cycling.
  - Frosts.
- Cables, breakers and fuses:
  - At least 156% larger than the short circuit solar photovoltaic current.
  - Increase current as needed for albedo effects.
  - Never exceed the interrupt ratings of the fuses and breakers.
  - Current carrying conductors: Black cable is live conductor, white cable is grounded conductor.
  - Non current carrying equipment ground conductor is either bare copper or green cable.
- Use metal enclosures for effective heat dissipation from the electrical equipment.
- Ensure that the metal types used in the mounting system do not cause galvanic corrosion.
- Use stainless steel fasteners for longevity and corrosion resistance.

AC Component Selection

- Always make sure that AC rated components are used.
- Components must exceed system specifications for maximum voltage.
- Components must exceed system specifications for maximum current.
- All outdoor cables must be solar rated.
- All outdoor connectors must be solar rated.
- De-rate equipment for installed conditions such as:
  - High ambient enclosure temperatures.
  - Burial.
  - Enclosed conduits and ducts.
  - Heating effects.
  - Cooling effects.
  - High power cycling.
- Cables, breakers and fuses:
  - At least 125% larger than the continuous AC current.
  - Apply de-rating for highest expected fuse ambient temperature per fuse manufacturers de-rating tables.
  - Never exceed the interrupt ratings of the fuses and breakers.
  - Use metal enclosures for effective heat dissipation.

Enclosure Mounting

- Exterior equipment locations should always be in the shade.

Inverters

- More inverters improves maximum power point tracking on the system
- Keep inverter power down to about 250 kW AC maximum per single inverter.
- Try to keep below 100 strings per inverter.
- If possible, mount inverters in shaded locations, consider constructing a shade canopy if needed.
- If inside a building or structure, ensure that the indoor ambient temperature can never exceed the inverter ambient temperature ratings.
- Use proven inverter technology for your system size.

Fencing

- Make sure that fencing does not shade the array.
- Make sure that the fencing is electrically grounded.
- Install access points that are sufficiently sized for the equipment that will pass through them.

Complete Solar Photovoltaics © Steven Magee

Manufacturer Instructions

- Always design your system in accordance with the manufacturers installation manuals.
- Follow the maintenance schedules in order to maintain warranty coverage.

Codes

- Always design to local photovoltaic electrical codes.
- Always design to local building codes.
- Consult with qualified engineers.
- Have qualified engineers approve designs.
- When in doubt, engineer on the side of safety.

Wildlife

- Use rodent barriers and screens.
- Install bird spikes.

Labeling

- Make sure all equipment is labeled in accordance with the local codes and the manufacturers instructions.

- Use engraved metal labels to avoid sunlight aging effects.

Safety

- Make sure all system and equipment grounding is installed correctly.
- Make sure all covers are installed.
- Make sure DC polarity of cabling is correct.
- Make sure AC phase rotation is correct for three phase systems.
- Follow the local safety codes.

System Inefficiencies

The system inefficiencies comprise of the following component de-rate factors:

- PV module nameplate DC rating 0.80 - 1.05
- Inverter and Transformer 0.88 – 0.98
- Mismatch 0.97 - 0.995
- Diodes and connections 0.99 – 0.997
- DC wiring 0.97 - 0.99
- AC wiring 0.98 – 0.993
- Soiling 0.30 - 0.995
- System availability 0.00 - 0.995
- Shading 0.00 – 1.00
- Sun-tracking 0.95 – 1.00
- Age 0.70 - 1.00

A well built solar photovoltaic system should be in the vicinity of 78% to 85% overall efficiency conversion from DC to AC at the point of interconnection (POI). If installed in an area with a high level of dirt buildup, then you may find that you are operating much lower than this and efficiencies in the 65% range are not unusual.

System size

There are two ways that are traditionally used to quote system sizes:

- Installed solar module DC power at STC.
- Expected AC output at system interconnection at STC.

It is important when designing a system that you quote both values accurately to the customer.

Grid Interconnection Requirements

Grid interconnection is different all over the world and it will depend on the grid you are connecting into. Each utility will have its own requirements for grid interconnection and you will need to contact them to obtain their documentation on this.

In general they are looking for the following attributes:

- Address of installed system.
- System generation capacity.
- Power factor of system generation.
- Inverter manufacturer(s) and model number(s).
- Sufficient interrupt ratings on breakers and fuses.
- For residential and commercial installations only, confirmation of anti-islanding feature.
- Certification of country standards of manufacture of equipment.

It is important as a utility customer that you use equipment that will disconnect upon the grid power failing. This is to ensure that the grid is not back fed and energized from your equipment. This represents a significant hazard to utility line workers and many have been exposed to electrical hazards caused by this effect in the past. UL 1741 listed inverter equipment meets this requirement.

For a large utility scale system, you will want the utility to supply the following details:

- Interconnection fault current availability.
- Interconnection voltage.
- Transformer configuration.
- Relay settings.

With these values you will be able to design your utility interconnection.

In areas that have a large amount of interconnected power generation systems, the utility will need to do a study of their system to ensure that instability will not be induced into their grid by the introduction of a large amount of power into it. This is called "high penetration" and typically may become an issue when the power on the grid reaches over 10% of interconnected power systems. The larger the penetration, the more significant the effects become. Typically over voltage may occur, harmonics, and large power swings that match the solar irradiance levels. The electrical grid may collapse if the relaying has not been set correctly for the effects from interconnected power systems.

On net metering installations, you will need to install a meter socket on the solar power system. Net metering means that the system will have a utility meter and a separate solar power meter. The utility will adjust your billing based on the two meter readings.

Complete Solar Photovoltaics © Steven Magee

Solar Power System Meter

You should check whether you need to install a meter for the solar power system. Sometimes it is required and other times it is not.

Complete Solar Photovoltaics © Steven Magee

On residential and commercial systems you have the choice of a load or supply side interconnection. Load side interconnections need to have a spare breaker at the end of the switchgear busbar per NEC 690.64(B). If the main breaker for the switchgear is rated for bi-directional current flow, then you can connect 20% of the main breaker rating to the switchgear. After applying the 125% de-rating factor for breakers, this comes out to 16% continuous current can be fed into the bus bar on the switchgear. Note that this only can be applied to the last breaker location.

For example, most homes have a 200A service. If you connected in a solar power system to the fuse board in the last breaker location, you would be limited to a maximum of a 40A breaker that has 32A continuous current flowing through it. The breaker must be installed in the last breaker location at the opposite end of the bus-bar from the main 200A circuit breaker for the fuse board.

When the fuse board is not large enough for solar, you should consider replacing it with one that is designed for solar power system use, or a larger fuse board. It may be a better route if the electrical fuse board is old.

Supply side connections are generally cleaner on residential and commercial installations and remove the need to be limited in size by the switchgear. However, you will need to coordinate with the utility in order to effect this.

Many inverters now come with both the DC and AC disconnects built into them. Some utilities still require

Complete Solar Photovoltaics © Steven Magee

you to install separate AC and DC disconnects for inverter maintenance purposes. You should verify with the utility what their requirements are for this provision.

Electrical Codes

The National Electrical Code (NEC) is published every three years and section 690 applies to solar photovoltaics. You should familiarize yourself with this section of the NEC prior to designing solar photovoltaic systems. Solar photovoltaics is a rapidly developing field and this section of the code changes with every edition. It is good practice to reference the current edition, even if you are not required to use it in your design. Most planning departments generally use an older edition for their standard.

NEC section 690 is a rich resource of technical information. Section 690 references many other sections of the NEC and some of these are:

240: Overcurrent Protection.

250: Grounding and Bonding.

285: Surge Protective Devices (SPD), 1kV or Less.

310: Conductors for General Wiring.

705: Interconnected Electric Power Production Sources.

Naturally, if you are designing solar photovoltaic systems, you should be adept in using the NEC. Make sure you are familiar with the contents of the book and perhaps consider reading chapter 1 again to refresh on the basics of the NEC.

Complete Solar Photovoltaics © Steven Magee

The following items in the NEC will be frequently referred to by the solar photovoltaic designer:

Table 250.122 Minimum Size Equipment Grounding Conductors for Grounding Raceway and Equipment.

250.166 Size of the Direct Current Grounding Electrode Conductor.

Table 310.15(B)(2)(a) Adjustment Factors for More Than Three Current Carrying Conductors in a Raceway or Cable.

Table 310.15(B)(2)(c) Ambient Temperature Adjustment for Conduits Exposed to Sunlight on or Above Rooftops.

Table 310.16 Allowable Ampacities of Insulated Conductors Rated 0 Through 2,000 Volts, 60 °C Through 90 °C (140 °F Through 194 °F), Not More Than Three Current-Carrying Conductors in Raceway, Cable, or Earth (Directly Buried), Based on Ambient Temperature of 30 °C (86 °F).

Table 310.17 Allowable Ampacities of Single-Insulated Conductors Rated 0 Through 2,000 Volts in Free Air, Based on Ambient Air Temperature of 30 °C (86 °F).

690.47 Grounding Electrode Systems.

Note that the DC and AC grounding circuits are bonded together for solar power systems.

Chapter 9, Table 8 Conductor Properties.

Chapter 9, Table 9 Alternating-Current Resistance and Reactance for 600-Volt Cables, 3 Phase, 60 Hz, 75 °C (167 °F) – Three Single Conductors in Conduit.

System Sizing

Sizing your system is a complicated equation of your energy demands and the area that you install your system.

"Net zero" is a common term in the solar industry. This means that your solar photovoltaic system is sized to match your annual consumption of electricity from the grid. In other words you use no electricity from the grid during the year when averaged out over the year. If you are looking for a net zero system, then the first step is to find out how much electricity you use during the year from your electricity bills.

Divide the annual energy consumption by 365 days to get the daily figure.

The next step is to look at the annual solar radiation charts for your area. These can be obtained at the National Renewable Energy Laboratories website at www.nrel.gov. The NREL USA chart for annual average solar data for photovoltaic systems is shown on the next page and was obtained from http://www.nrel.gov/gis/solar.html. I would recommend that you take a look at it in color at the website and download it. You can find out an estimate of how many hours of average daily solar radiation that your area will get from these charts.

Solar photovoltaic system size will be the daily energy consumption divided by the daily solar radiation.

Complete Solar Photovoltaics © Steven Magee

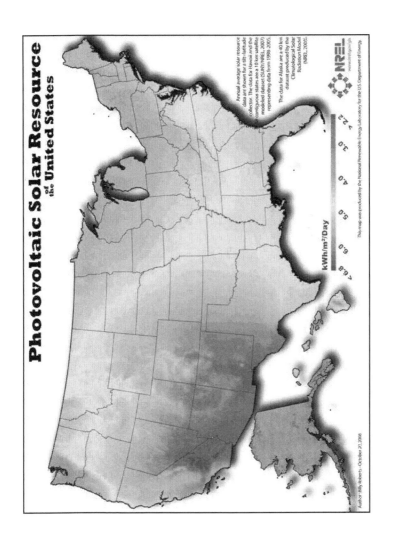

Complete Solar Photovoltaics © Steven Magee

Using the commonly used figure of 0.77 DC to AC derating for the solar power system, adjust the system size figure for the photovoltaic system efficiency losses.

You can adjust for aging effects using a 1% per year of loss of efficiency.

For example:

This particular sample household uses 7,300 kilowatt hours (kWh) of electrical energy per year.

The system that will be installed will be a South facing fixed tilt system that is inclined to match the latitude angle. There is no solar photovoltaic module shading during the day.

Divide 7,300 kWh per year by the number of days per year

= 7,300 / 365

= 20 kWh per day

Looking at the solar radiation charts, this area gets an an annual average of 5 kilowatt hours per square meter per day.

Solar radiation = 5 kWh/m^2/Day

Complete Solar Photovoltaics © Steven Magee

We now divide the daily household power requirements by the number of hours of solar energy for this area.

20 kWh / 5 hours

= 4 kW AC solar photovoltaic system size

We need to adjust this for the solar photovoltaic system DC to AC derating of 0.77 to get the correct DC system size.

= 4 kW / 0.77 de-rating

= 5.19 kWp DC solar photovoltaic system size

Since solar photovoltaic power systems do suffer from aging effects by about 1% per year reduction in power from new, we will factor this in at a 10% loss after 10 years of service and increase the system size accordingly to compensate.

5.19kW / 0.9

5.77 kWp DC

So our system should produce more energy when new and after ten years will hit the net zero figure as it ages.

Complete Solar Photovoltaics © Steven Magee

It is always good practice to oversize solar power systems so that the owner is not disappointed by poor performance. There is nothing worse than a system that does not meet its minimum performance expectations, which in this case is a net zero system.

A further adjustment is required for tracking systems:

A single axis tracker that is inclined at the latitude angle can produce an increase of energy production by about a factor of 1.2. We will reduce the DC system in size accordingly which gives:

= 5.77 kWp DC / 1.2

= 4.81 kWp DC single axis tracker system size

A dual axis tracker can increase the power production by about a factor of 1.3. We will reduce the system size accordingly:

= 5.77 kWp DC / 1.3

= 4.44 kWp DC dual axis tracker system size

You must remember that solar photovoltaics is more of an art than a science and these figures are approximations. It is common for some systems to over perform and others to under perform. It really just depends on your local conditions.

The safest way to quote solar photovoltaic power generation systems is installed DC capacity and installed AC capacity after conversion losses to the point of interconnection with the grid at STC. Annual power production predictions can be unreliable at times and should not be relied on.

Most companies when building a large power plant will build in expansion to the design at the start of the project as an insurance policy in case the system under performs in its location. Generally, a 10% expansion buffer is sensible to use for this purpose. It is wise to oversize the system to exceed the desired annual power production figure and to build it bigger if it under performs once in operation.

The above calculations are based on averaged data from NREL. For a large solar site, they should not be a substitute for a on site study of climatic and system annual performance for at least one year in duration prior to design and construction of a project. It is good to obtain several years of data if possible for a solar photovoltaic site prior to development.

The NREL website has a huge amount of historical solar radiation information on it and it is well worth taking the time to check out their latest information. The NREL data should be the starting point for planning a large solar photovoltaic project.

Sizing for power production for a certain month of the year is occasionally desired. For example, a common request is that a utility will want the solar photovoltaic power plant to cover the peak summertime load of the air conditioning loads on their system. Usually this type of load peaks in July.

To do this we look at the monthly solar radiation NREL charts and this is shown on the next page for July. It was downloaded from http://www.nrel.gov/gis/solar.html. I would recommend that you take a look at it in color at the website and download it. You can find out an estimate of how many hours of average daily solar radiation that your area will get over one month from these charts.

Using this chart, this area gets an average of 6 kilowatt hours (kWh) of solar radiation (called insolation) per square meter per day in July.

Solar radiation = 6 kWh/m^2/Day

We run the calculations, however, this time we factor in the temperature of the system as well as it will be hot. The peak ambient temperature for this area is 40 degrees Celcius and we will assume an average solar module surface temperature of 80 degrees for our calculations.

The utility wants to generate a daily average of 10 megawatts hours (MWh) of energy with the system in July.

Daily energy production = 10 MWh per Day

Complete Solar Photovoltaics © Steven Magee

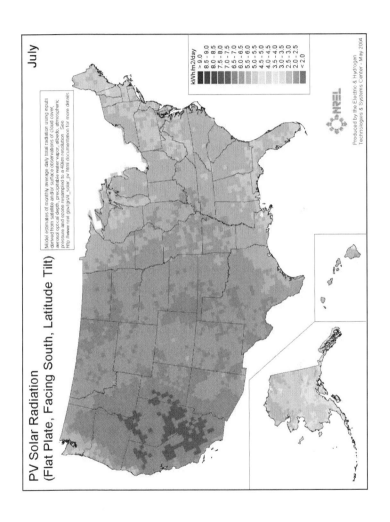

The system that will be installed will be a South facing fixed tilt system that is inclined to match the latitude angle. There is no solar photovoltaic module shading during the day.

We now divide the daily power requirements by the number of hours of solar energy for this area.

= 10MWh / 6

= 1.67 MW AC solar photovoltaic system size

We need to adjust this for the solar system DC to AC derate factor of 0.77 to get the correct DC system size.

= 1.67 MW / 0.77

= 2.16 MWp DC solar photovoltaic system size

Since solar power systems do suffer from aging effects by about 1% per year from new, we will factor this in for 10 years of service and increase the system size accordingly to compensate.

= 2.16 MWp / 0.9

= 2.4 MWp DC system size

Finally we apply the temperature correction for an 80 degrees Celsius solar module temperature which is -0.485% for power per degree on a silicon solar module, so we correspondingly increase the system size:

= System DC power + (system DC power x STC temperature difference x solar module power coefficient)

= 2.4 + (2.4 x 0.485% x 55)

= 3.04 MWp DC system size

So our system should produce more energy when new and after ten years it may hit the desired July 10 MWh figure as it ages.

It is always good practice to oversize solar power systems so that the owner is not disappointed by poor performance. There is nothing worse than a system that does not meet its minimum performance expectations.

A further adjustment is required for tracking systems:

A single axis tracker that is inclined at the latitude angle produces an increase of energy production by about a factor of 1.2. We will reduce the DC system in size accordingly which gives:

= 3.04 MWp DC / 1.2

= 2.53 MWp DC system size for inclined single axis tracker

A dual axis tracker can increase the daily power production by about a factor of 1.3. We will reduce the system size accordingly:

= 3.04 MWp DC / 1.3

= 2.34 MWp DC system size for a dual axis tracker

The utility may request that this system must produce a minimum amount of energy in summertime of 10MW on cloudy days. We would simply adjust our figures to account for this with an increase of system size to match the decrease in solar energy. For this example, the utility has requested 10 MW per day in irradiance conditions of 250 W/m^2. We have only 25% approximately of our sunlight, assuming July is a sunny month with clear skies, so we correspondingly increase the system size to account for this.

250W/m^2 10MWp Fixed tilt system:

= 3.04 MWp / 0.25

= 12.16 MWp DC system

250 W/m^2 10MW inclined single axis tracker

= 12.16 MWp / 1.2

= 10.13 MWp system size

250W/m^2 10MW Dual axis tracker

= 12.16 MWp /1.3

= 9.35 MWp DC system size

As you can see we have used a inverse approximation between irradiance and system size. As irradiance decreases our system size correspondingly needs to increase in size to keep delivering the same output power. This can be further refined by looking at the power output for the solar module technology selected for your system and using the actual power listed for 250W/m^2 irradiation levels.

Now that we have a good understanding of how to perform solar photovoltaic system sizing calculations, let me introduce you to the easy way of doing this. Most solar photovoltaic system designers will use the PVWatt calculators at the NREL website http://www.nrel.gov/rredc/pvwatts/ to size their systems.

You must remember that solar photovoltaics is more of an art than a science and these figures are approximations. It is common for some systems to over perform and others to under perform. It really just depends on your local conditions.

Solar Modules

Solar modules typically represent the largest investment in the system. Traditionally silicon solar modules have been used, but now many new types of technologies are emerging such as thin film and so on. All solar modules should be tested to the Underwriters Laboratory (UL) 1703 standard or equivalent in the USA.

In the following chapters we will take a look at each technology and see the needs of each technology when building residential, commercial and utility systems with them. The main solar technologies in use currently are:

- Mono-crystalline silicon wafers (also called single crystal)
- Poly-crystalline silicon wafers (also called multi-crystalline)
- Amorphous silicon thin film
- Cadmium telluride thin film
- CIGS thin film

Crystalline silicon is the best understood, the most efficient and has been around for many decades. The newer film type solar modules are less efficient and cheaper to purchase. Unfortunately, more thin film modules are needed to generate the same power as mono and poly crystalline modules, and this increases the system physical size and associated support systems such as land purchase, cabling, racking, installation costs, operation and maintenance costs, and so on. Thin film uses less materials to produce it and it really is a film, usually about 1/50 the thickness of a silicon wafer solar cell.

Thin film is a relatively new technology in the solar photovoltaic field. Development of thin film systems was in process in the late nineties and became a commercial product in the last decade. Thin film benefited from a surge in silicon wafer costs and a lack of supply of silicon solar modules in 2000-2010 period. Thin film was widely installed during this time. Recently, this silicon supply problem has been rectified and silicon prices have fallen sharply and this has been hurting the thin film solar module manufacturers due to the superior efficiency of silicon wafers.

A further problem with thin film is that its longevity is still being assessed. Some thin film manufacturers offer a 25 year warranty but cannot demonstrate commercial product that is of that age. It will come with time, but financing companies are reluctant to finance thin film due to the higher risks that aging poses for these systems. In time, the various thin film solar modules are expected to be proven to have a similar reliability as silicon wafer solar modules.

Thin film systems are generally marketed as the solar module of choice in cloudy locations due to their improved performance in low light conditions when compared to silicon. Thin film is less efficient than silicon and is currently between 7 and 12 percent in the commercial products. As such more solar modules are needed when compared to crystalline silicon systems to generate the same energy at STC.

Generally any decision on which technology to use is driven by market rates for each type of technology, aesthetics and personal preference. Solar modules are a commodity and their prices can fluctuate rapidly. When designing a system, it may be built from any of the above mentioned solar modules.

Roofing materials are now being manufactured with solar photovoltaics integrated into them. This is a great idea and will bring down the installation costs on new construction. The products are relatively new and it will be interesting to watch this area develop. Generally this technology is based on thin film.

Be careful when using manufacturers data sheets for their solar modules. The module power value generally has percentage tolerances for negative and positive adjustments to this figure. This is a reflection of the imprecision of the solar photovoltaic cell manufacturing processes. Due to this it is common for individual strings of solar modules to either over or under perform in the system. For example, the cloud effect combined with a 10% high performing string of solar modules could produce 165% more than the rated STC current from the string. If your system under performs it may be due to having the bulk of the supplied modules performing at lower than normal power ratings. At least one major manufacturer does not list any negative adjustments on its solar photovoltaic module data sheets, only positive adjustments. This greatly helps with meeting the system performance requirements. It would be good to see other manufacturers follow their lead.

There are some manufacturers that offer 25 year warranties but do not have any product that old. You will need to assess if this is an issue in your decision to use their products. A good question would be to ask yourself if you think that the manufacturer will be in existence in 25 years time to honor the warranty.

Electrically, a solar module is very simple device. It turns light into current at a rate that is proportional to the irradiance it receives. It does this at a voltage that is dependent on the temperature of the solar cells. The solar module will keep indefinitely producing current proportional to the levels of irradiance it receives, even if these levels are much higher than STC due to reflections and lensing. The solar module does this conversion in real time, there is no delay in the conversion process.

A simple approximation of a solar module is that it is a current limited source and that the upper limit for current is entirely dependent on the amount of light received by it!

Due to this, it pays to use the maximum fuse size as recommended by the manufacturer of the solar module.

If the irradiance levels are very low at either dusk or dawn then the voltage will be somewhere between 0V and Voc when adjusted for solar cell temperature. At night time it will be at or very close to 0V. The inverter system that the solar photovoltaic modules are connected to knows when to turn on and turn off by monitoring the voltage rise at dawn and the voltage fall at dusk. The inverter system will generally switch on and off at preset voltage levels that have been entered into it during installation.

Solar modules commonly have what is known as "bypass diodes" installed in them. These are installed for shading effects. This enables the other modules in the string to function if one is shaded.

Less common on grid tied systems are "blocking" diodes. One per string can be installed and this prevents reverse currents from flowing into the strings from the system. The string fuses now protect against this by blowing if the reverse current gets too high in a grid tied solar photovoltaic system. It is for this reason that the manufacturer limits the maximum size of the string fuse. If the fuse is above the size recommended and reverse currents occur, then the solar modules in the string may be damaged beyond repair.

If you take a look at the solar module data sheets from the various manufacturers you will see graphs showing the above concepts for their products. Take time to look at the data sheets and understand them, they have a lot of information in them. Don't build a system with their product until you have read and fully understood the data sheet for it.

We will look into the physical sizes of three systems for each solar module type. These will be:

Complete Solar Photovoltaics © Steven Magee

- 5 kWp DC typical residential system
- 250 kWp DC typical commercial system
- 10 MWp DC typical utility system

By comparing the system sizes and components you will be able to see the advantages of each of the different mainstream technologies.

We will not look at the newer solar photovoltaic technologies due to them being in their infancy. Their longevity is currently unknown and most large financing companies will not finance these newer technologies in large installations.

The following picture shows the solar module and the inside of the junction box on the rear of the module. The circuit diagram shows how a crystalline silicon solar photovoltaic module is wired together. As you can see, it is a collection of very large solar cells wired together into strings. The bypass diodes electrically bypass the internal strings of solar cells if they are either faulty or shaded.

The picture on the following page demonstrates the string concept of solar modules. Solar modules are first wired in series, this is called "stringing", to increase their DC voltage output. Then many strings are connected in parallel to increase the current output of the solar system. Raising the voltage and current is needed for the inverter system to work properly and to increase the efficiency of the system.

Let us now take a look at each of these technologies and see the differences between typical systems of the same size that are built using the different technologies.

Complete Solar Photovoltaics © Steven Magee

A Two String Mono-Crystalline Solar Module

Inside the Solar Module Junction Box

Note the two bypass diodes.

INTERNAL DIAGRAM OF A SOLAR MODULE

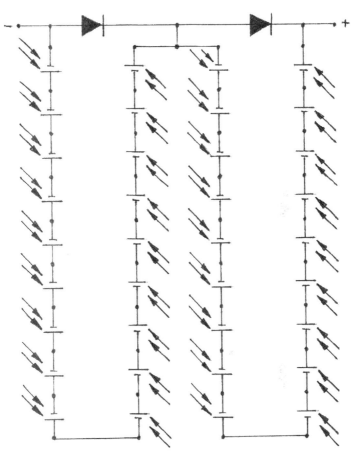

A CRYSTALLINE SILICON SOLAR MODULE IS MADE OF LARGE SOLAR CELLS AND BYPASS DIODES.

© COPYRIGHT STEVEN MAGEE

Complete Solar Photovoltaics © Steven Magee

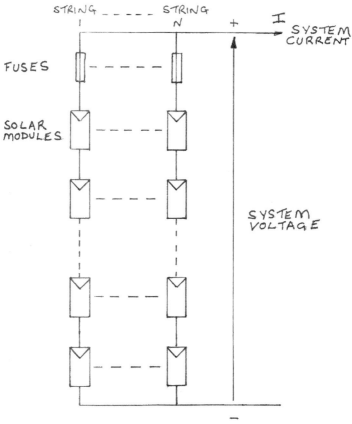

Mono-Crystalline Modules

Mono-crystalline modules are the highest efficiency solar modules available in the general market. What does this mean? Basically an installation constructed with these modules will use less support equipment as these systems use the least amount of solar modules for a given energy requirement. This advantage comes at a cost, as these modules are also the most expensive.

They are made out of a high grade silicon wafers and are very reliable. The mono-crystalline solar cells look black to the eye. Mono-crystalline solar modules have been around for many decades and are very well understood and commonly come with 25 year warranties. Most major manufacturers can show you modules that exceed this age and that are still functioning well today.

Mono-crystalline from most manufacturers is in the 14 to 18 percent efficiency rating. A few have actually manage to exceed 18 percent with over 20 percent being achieved commercially.

Typical Mono-crystalline Solar Module Specifications

- 600V maximum system voltage
- 235W power rating +10% / -5%
- 37V open circuit (Voc)
- 8.6A short circuit (Isc)

- 30V maximum power point (Vmpp)
- 7.84A maximum power point (Impp)
- -0.351% voltage (Voc) temperature coefficient
- +0.053% current (Isc) temperature coefficient
- 15A maximum fuse
- Length 1.64m / 64.6"
- Width 0.994m / 39.1"
- 1.63m^2 / 17.53 sq ft module surface area

System Calculations

Looking at these dimensional values we get a square meter power of:

W/m^2 = rated power / surface area

= 235W / 1.63m^2

= 144.17 W/m^2

We dived this value by ten to get the module efficiency per square meter:

Module efficiency = 14.417% per m^2

Assuming a lowest ambient temperature of -10 °C for this area, we need to calculate the maximum voltage for this solar module:

Vmax = (Difference from STC temperature x Voc temperature coefficient x Voc) + Voc

= ((-35) x (-0.351%) x 37V) + 37V

= 41.55V maximum adjusted solar module voltage

Maximum number of solar modules in the strings:

= Rated system voltage / maximum adjusted solar module voltage

= 600V / 41.55

= 14 solar modules maximum in the strings

Wattage of maximum individual string:

= # modules x module wattage

= 14 x 235W

= 3,290W per string

Complete Solar Photovoltaics © Steven Magee

5kWp DC System

Number of modules required in the 5kWp system:

= System wattage / module wattage

= 5,000W / 235W

= 22 Modules

\# strings = # system modules / # modules in string

= 22 / 14

= 2 strings, each of 11 modules

5kWp system surface area:

= # modules x module surface area

= 22 x 17.53 sq ft

= 385.66 sq ft module surface area

250 kWp DC system

Number of modules required in the 250 kWp system:

= System wattage / module wattage

= 250,000W / 235W

= 1,064 Modules

strings = # system modules / # modules in string

= 1,064 / 14

= 76 strings

Number of 12 input combiner boxes required:

= # strings / # combiner box inputs

= 76 / 12

= 7 combiner boxes required

250kWp system surface area:

= # modules x module surface area

= 1,064 x 17.53 sq ft

= 18,651.92 sq ft module surface area

10 MWp DC system

Number of modules required in the 10MWp system:

= system wattage / module wattage

= 10,000,000W / 235W

= 42,554 rounded up to 42,560 Modules

strings = # system modules / # modules in string

= 42,560 / 14

= 3,040 strings

Number of 12 input combiner boxes required:

= # strings / # combiner box inputs

= 3,040 / 12

= 254 combiner boxes required

10MWp system surface area:

= # modules x module surface area

= 42,560 x 17.53 sq ft

= 746,076.8 sq ft module surface area

Let's make an assumption that when constructed and allowing for spacing between the modules for shading effects that this system will cover 100 acres. We will compare the sizes of our other technologies to this to develop the size of land needed for each technology type.

Poly-Crystalline Modules

Poly-crystalline solar modules are made from lower grade and therefore cheaper silicon. As such, their efficiency is less than mono-crystalline, but not by that much. A poly-crystalline solar module has a typical conversion efficiency of 13 percent to 16 percent. Their purchase price reflects this lower efficiency and the dollar per watt figure is typically less than mono-crystalline. Poly-crystalline solar cells typically look a metallic blue color to the eye. They are visually a very attractive module to the eye.

Poly-crystalline solar modules have been around for many decades and are a very stable and proven technology.

Typical Poly-crystalline Solar Module Specifications

- 600V maximum system voltage
- 215W power rating +5W / 0W
- 33.2V open circuit (Voc)
- 8.78A short circuit (Isc)
- 26.6V maximum power point (Vmpp)
- 8.09A maximum power point (Impp)
- -0.361% voltage (Voc) temperature coefficient
- +0.06% Current (Isc) temperature coefficient
- 15A maximum fuse
- Length 1.5m / 59.1"

- Width 0.99m / 39"
- 1.485m² / 16 sq ft module surface area

System Calculations

Looking at these dimensional values we get a square meter power of:

W/m^2 = rated power / surface area

= 215W / 1.485m²

= 144.78 W/m^2

We dived this value by ten to get the module efficiency per square meter:

Module efficiency = 14.478% per m^2

Assuming a lowest ambient temperature of -10 °C for this solar site, we need to calculate the maximum voltage for this solar module:

Vmax = (Difference from STC temperature x Voc temperature coefficient x Voc) + Voc

= ((-35) x (-0.361%) x 33.2V) + 33.2V

= 37.395V maximum adjusted solar module voltage

Maximum number of solar modules in the strings:

= Rated system voltage / maximum adjusted solar module voltage

= 600V / 37.395

= 16 solar modules maximum in the strings

Wattage of maximum individual string:

= # modules x module wattage

= 16 x 215W

= 3,440W per string

5kWp DC System

Number of modules required in the 5kWp system:

= System wattage / module wattage

= 5,000W / 215W

= 24 Modules

\# strings = \# system modules / \# modules in string

= 24 / 16

= 2 strings, each of 12 modules

5kWp system surface area:

= \# modules x module surface area

= 24 x 16 sq ft

= 384 sq ft module surface area

250 kWp DC system

Number of modules required in the 250 kWp system:

= System wattage / module wattage

= 250,000W / 215W

= 1,168 Modules

\# strings = \# system modules / \# modules in string

= 1,168 / 16

= 73 strings

Number of 12 input combiner boxes required:

= # strings / # combiner box inputs

= 73 / 12

= 7 combiner boxes required

250kWp system surface area:

= # modules x module surface area

= 1,168 x 16 sq ft

= 18,688 sq ft module surface area

10 MWp DC system

Number of modules required in the 10MWp system:

= system wattage / module wattage

= 10,000,000W / 215W

= 46,512 modules

strings = # system modules / # modules in string

= 46,512 / 16

= 2,907 strings

Number of 12 input combiner boxes required:

= # strings / # combiner box inputs

= 2,907 / 12

= 243 combiner boxes required

10MWp system surface area:

= # modules x module surface area

= 42,560 x 17.53 sq ft

= 744,192 sq ft module surface area

Comparing this to our mono-crystalline system we get:

Land size = 100 acres x poly surface area / mono surface area

= 100 x 744,192 / 746,078.8

= 99.74 acres

As we can see, there is no significant difference between the area for the mono and poly crystalline modules in this application.

Amorphous Silicon (a-Si) / Micro-Crystalline Silicon Thin Film

Commonly referred to as the calculator solar cell, due to its use in small solar powered consumer devices. This material has been around for many decades and its properties are well known and understood.

Typical Amorphous Silicon / Micro-Crystalline Silicon Solar Module Specifications

- 600V maximum system voltage
- 135W power rating
- 249V open circuit (Voc)
- 0.87A short circuit (Isc)
- 188V maximum power point (Vmpp)
- 0.72A maximum power point (Impp)
- -0.3% voltage (Voc) temperature coefficient
- +0.07% Current (Isc) temperature coefficient
- 2A Fuse
- Length 1.409m / 55.5"
- Width 1.009m / 39.7"
- $1.422m^2$ / 15.29 sq ft module surface area

System Calculations

Looking at these dimensional values we get a square meter power of:

W/m^2 = rated power / surface area

= 135W / 1.422m^2

= 94.94 W/m^2

We divide this value by ten to get the module efficiency per square meter:

Module efficiency = 9.494% per m^2

Assuming a lowest ambient temperature of -10 °C for this solar site, we need to calculate the maximum voltage for this solar module:

Vmax = (Difference from STC temperature x Voc temperature coefficient x Voc) + Voc

= ((-35) x (-0.3%) x 249V) + 249V

= 275.2V maximum adjusted solar module voltage

Maximum number of solar modules in the strings:

= Rated system voltage / maximum adjusted solar module voltage

= 600V / 275.2V

= 2 solar modules maximum in the strings

Wattage of maximum individual string:

= # modules x module wattage

= 2 x 135W

= 270W per string

5kWp DC System

Number of modules required in the 5kWp system:

= System wattage / module wattage

= 5,000W / 135W

= 37 rounded to 38 modules

strings = # system modules / # modules in string

= 38 / 2

= 19 strings

5kWp system surface area:

= # modules x module surface area

= 38 x 15.29 sq ft

= 581.02 sq ft module surface area

250 kWp DC system

Number of modules required in the 250 kWp system:

= System wattage / module wattage

= 250,000W / 135W

= 1,852 Modules

strings = # system modules / # modules in string

= 1,852 / 2

= 926 strings

Number of 12 input combiner boxes required:

= # strings / # combiner box inputs

= 926 / 12

= 78 combiner boxes required

250kWp system surface area:

= # modules x module surface area

= 1852 x 15.29 sq ft

= 28,317.08 sq ft module surface area

10 MWp DC system

Number of modules required in the 10MWp system:

= system wattage / module wattage

= 10,000,000W / 135W

= 74,075 rounded up to 74,076 modules

\# strings = # system modules / # modules in string

= 74,076 / 2

= 37,038 strings

Number of 12 input combiner boxes required:

= # strings / # combiner box inputs

= 37,038 / 12

= 3,087 combiner boxes required

10MWp system surface area:

= # modules x module surface area

= 74,075 x 15.29 sq ft

= 1,132,606.75 sq ft module surface area

Comparing this to our mono-crystalline system we get:

Land size = 100 acres x silicon film surface area / mono surface area

= 100 x 1,132,606.75 / 746,078.8

= 151.818 acres

As we can see, there is a significant difference between the land requirements of the silicon crystalline modules and the silicon film in this application. At this point you would be wise to calculate the initial land costs, installation costs, and the expected operation and maintenance costs for the lifetime of the system.

Cadmium Telluride (CdTe) Thin Film

This is the market leader in thin film. It is a coating process that is applied directly to the glass. A nice feature of it is that it does not need bypass diodes, so the solar module junction box is much simpler. This is a relatively young technology and is in widespread use and well understood. It has better high temperature performance than the other technologies.

Typical Cadmium Telluride Solar Module Specifications

- 600V maximum system voltage
- 70W power rating
- 88V open circuit (Voc)
- 1.23A short circuit (Isc)
- 65.5V maximum power point (Vmpp)
- 1.07A maximum power point (Impp)
- -0.25% voltage (Voc) high temperature coefficient
- -0.2% voltage (Voc) low temperature coefficient
- +0.04% Current (Isc) temperature coefficient
- 10A Fuse
- Length 1.2m / 47.244"
- Width 0.6m / 23.622"
- $0.72m^2$ / 7.745 sq ft module surface area

System Calculations

Looking at these dimensional values we get a square meter power of:

W/m^2 = rated power / surface area

= 70W / 0.72m^2

= 97.22 W/m^2

We divide this value by ten to get the module efficiency per square meter:

Module Efficiency = 9.722% per m^2

Assuming a lowest ambient temperature of -10 °C for this solar site, we need to calculate the maximum voltage for this solar module:

Vmax = (Difference from STC temperature x Voc temperature coefficient x Voc) + Voc

= ((-35) x (-0.2%) x 88V) + 88V

= 94.16V maximum adjusted solar module voltage

Maximum number of solar modules in the strings:

= Rated system voltage / maximum adjusted solar module voltage

= 600V / 94.16V

= 6 solar modules maximum in the strings

Wattage of maximum individual string:

= # modules x module wattage

= 6 x 70W

= 420W per string

5kWp DC System

Number of modules required in the 5kWp system:

= System wattage / module wattage

= 5,000W / 70W

= 72 modules

strings = # system modules / # modules in string

= 72 / 6

= 12 strings

5kWp system surface area:

= # modules x module surface area

= 72 x 7.745 sq ft

= 557.64 sq ft module surface area

<u>250 kWp DC system</u>

Number of modules required in the 250 kWp system:

= System wattage / module wattage

= 250,000W / 70W

= 3,572 Modules

strings = # system modules / # modules in string

= 3,572 / 6

= 596 strings

Number of 12 input combiner boxes required:

= # strings / # combiner box inputs

= 596 / 12

= 50 combiner boxes required

250kWp system surface area:

= # modules x module surface area

= 3572 x 7.745 sq ft

= 27,665.14 sq ft module surface area

10 MWp DC system

Number of modules required in the 10MWp system:

= system wattage / module wattage

= 10,000,000W / 70W

= 142,858 modules

\# strings = # system modules / # modules in string

= 142,858 / 6

= 23,810 strings

Number of 12 input combiner boxes required:

= # strings / # combiner box inputs

= 23,810 / 12

= 1,985 combiner boxes required

10MWp system surface area:

= # modules x module surface area

= 74,075 x 15.29 sq ft

= 1,106,427.47 sq ft module surface area

Comparing this to our mono-crystalline system we get:

Land size = 100 acres x CdTe film surface area / mono surface area

= 100 x 1,106,427.47 / 746,076.8 acres

= 148.3 acres

As we can see, there is significant difference in land use between the silicon crystalline modules and the CdTe film in this application. At this point you would be wise to calculate the initial land costs, installation costs, and the expected operation and maintenance costs for the lifetime of the system.

CIGS Thin Film

CIGS comprises of copper, indium, gallium, and selenium in its manufacturing process, hence its name. CIGS thin film looks a streaky greenish black to the eye due to the bath process that is used during the manufacture of the product. CIGS was developed in the nineties and has been a relatively new entrant to the solar photovoltaic field. It is a vacuum deposition process.

Typical CIGS Thin Film Solar Panel Specifications

- 600V maximum system voltage
- 150W power rating
- 96V open circuit (Voc)
- 2.5A short circuit (Isc)
- 70.5V maximum power point (Vmpp)
- 2.15A maximum power point (Impp)
- -0.38% voltage (Voc) temperature coefficient
- -0.06% Current (Isc) temperature coefficient
- 12A Fuse
- Length 1.82m / 71.65"
- Width 1.08m / 42.52"
- $1.97m^2$ / 20.65 sq ft module surface area

Complete Solar Photovoltaics © Steven Magee

System Calculations

Looking at these dimensional values we get a square meter power of:

W/m^2 = rated power / surface area

= 150W / 1.97m^2

= 76.14 W/m^2

We divide this value by ten to get the module efficiency per square meter:

Module efficiency = 7.61% per m^2

Assuming a lowest ambient temperature of -10 °C for this solar site, we need to calculate the maximum voltage for this solar module:

Vmax = (Difference from STC temperature x Voc temperature coefficient x Voc) + Voc

= ((-35) x (-0.38%) x 96V) + 96V

= 108.77V maximum adjusted solar module voltage

Maximum number of solar modules in the strings:

= Rated system voltage / maximum adjusted solar module voltage

= 600V / 108.77V

= 5 solar modules maximum in the strings

Wattage of maximum individual string:

= # modules x module wattage

= 5 x 150W

= 750W per string

5kWp DC System

Number of modules required in the 5kWp system:

= System wattage / module wattage

= 5,000W / 150W

= 34 rounded up to 35 modules

strings = # system modules / # modules in string

= 35 / 5

= 7 strings

5kWp system surface area:

= # modules x module surface area

= 35 x 20.65 sq ft

= 722.75 sq ft module surface area

250 kWp DC system

Number of modules required in the 250 kWp system:

= System wattage / module wattage

= 250,000W / 150W

= 1,667 rounded up to 1,670 Modules

strings = # system modules / # modules in string

= 1,667 / 5

= 334 strings

Number of 12 input combiner boxes required:

= # strings / # combiner box inputs

= 334 / 12

= 28 combiner boxes required

250kWp system surface area:

= # modules x module surface area

= 1,667 x 20.65 sq ft

= 34,423.55 sq ft module surface area

10 MWp DC system

Number of modules required in the 10MWp system:

= system wattage / module wattage

= 10,000,000W / 150W

= 66,667 rounded up to 66,670 modules

strings = # system modules / # modules in string

= 66,670 / 5

= 13,334 strings

Number of 12 input combiner boxes required:

= # strings / # combiner box inputs

= 13,334 / 12

= 1,112 combiner boxes required

10MWp system surface area:

= # modules x module surface area

= 66,667 x 20.65 sq ft

= 1,376,673.55 sq ft module surface area

Comparing this to our mono-crystalline system we get:

Land size = 100 acres x CIGS film surface area / mono surface area

= 100 x 1,376,673.55 / 746,076.8

= 184.52 acres

As we can see, there is significant difference between the land area for silicon crystalline modules and the CIGS film in this application. At this point you would be wise to calculate the initial land costs, installation costs and the expected operation and maintenance costs for the lifetime of the system.

Solar Module Summary

For residential installations, the supporting infrastructure costs vary little between the different technology types for installation, operation and maintenance.

This dynamic starts to change as the system gets larger with commercial and utility installations. As we can see, just an decrease in conversion efficiency of 1% can add acres of land use and supporting infrastructure to a 10 MWp project. The industry is constantly trying to find more efficiency out of their products due to this.

If land use is a significant project expense or is limited, then you probably should be using the more efficient technologies.

Some areas have very cheap land and if this is the case, the more important figures to look at are the supporting infrastructure costs for a larger installation and the ongoing annual operation and maintenance costs for a larger power generation facility. In the dry deserts of the South West USA, these operation and maintenance costs are generally quite low.

Thin film has better performance in low light conditions when compared to the silicon wafer technologies and this may be a factor to consider if the installation area is frequently cloudy.

If the failure rate is the same across all technologies, then the you will be replacing more equipment during the year on a less efficient installation during routine maintenance of the system. Manufacturers appear to be reluctant to release solar module failure rates for their solar module products and on any project, I have become accustomed to assuming a 1% annual failure rate in absence of this information. Unusual and unexpected severe weather conditions may increase this failure rate. Severe weather insurance is recommended to cover these events.

Operation and maintenance costs will be costly for some projects due to:

- Poor site selection
- Poor equipment selection
- Poor wildlife assessment
- Poor build quality

Performing due diligence at the start of any project will keep these problems under control. Ensuring that a highly skilled solar photovoltaic consultant is overseeing large projects from conception through to design and onto construction and commissioning will help to prevent these problems from occurring.

Site selection is probably the most important part of any large solar photovoltaic system project and can be the determining factor in a projects overall success. On any large project it is recommended to install a small version of the proposed system for the first year prior to constructing the main project so that an assessment can be made of how the system will perform in that location and design changes can be incorporated into the main project once all the risks are known. In the large scale solar photovoltaics field it is the tortoise that is generally more successful, try not to be the hare.

Be wary of locations close to the ocean and take the necessary precautions with equipment selection so that high winds, salt spray and corrosion don't become a problem. Salt will deposit itself onto the surface of the modules and if it is excessive, you will be frequently washing the solar photovoltaic modules. Expect the operation and maintenance costs to be higher when installing next to the ocean.

Frame-less solar photovoltaic modules are becoming available and these have better self cleaning properties during rains when compared to framed modules. Frame-less modules are not as strong as framed modules and you will need to assess if this is a problem for your installation or not. Some frame-less modules do not require a ground connection and this can make a significant saving on a large project.

The table on the next page shows the different technologies.

Complete Solar Photovoltaics © Steven Magee

Module Type	Module Wattage	Efficiency	Modules in Strings	# Modules	10 MW Area	Combiner Boxes
Mono	235W	14.42%	14	42,560	100 acres	254
Poly	215W	14.48%	16	46,512	99.74 acres	243
a-Si	135W	9.49%	2	74,076	151.8 acres	3,087
CdTe	70W	9.722%	6	142,858	148.3 acres	1,985
CIGS	150W	7.61%	5	66,667	184.5 acres	1,112

Inverters

Inverters convert the DC solar photovoltaic module power into AC power that can be put into the grid. Grid tied inverters come in many different types and sizes. This book is dedicated to grid tie inverters and this is what we will consider. All residential and commercial solar photovoltaic grid tie inverters should comply with UL1741 in the USA. This ensures reliability throughout the various climate zones that the USA has to offer.

There is a lot of discussion as to where inverter technology is going. There are cost savings with using the largest inverter available for large utility installations, but this comes at the expense of poorer MPPT tracking and long DC cable runs which cause power losses. The DC currents are very high in these systems at several thousand amps. Large solar photovoltaic inverter technology (>250 kW) is relatively new in the industry and may not be as reliable as the established small (<10kW) and medium (10kW to 250kW) sized inverters.

AC solar modules are starting to appear and this is at the smallest end of the inverter spectrum. Each solar module has it's own inverter attached to it, commonly called a "micro" inverter. This gives great MPPT performance. This is the ideal configuration for systems that may suffer from:

- Shading

- Have solar modules facing in different directions from each other
- Heavy dirt accumulation

Long term, as the "micro" inverter technology reduces in price I believe that this may become the standard in the industry, even on utility scale installations. The installation is simplified and removes DC from the circuit which can be problematic due to arcing effects which can cause equipment to burn out when it occurs.

Looking at today's technology I currently recommend conventional inverter systems for residential, commercial and utility scale installations. I do not recommend inverters over 250 kW as the DC currents and power flowing through these systems is very large and may cause problems in the solar DC circuit if it is not designed correctly.

Micro-inverters can be useful where there is nowhere to mount the traditional inverter on a domestic installation, or if the owner considers an inverter box ugly. The micro-inverters can be mounted to the solar modules and hidden from view.

The utility scale inverter systems are still in flux in the USA and currently there is no UL standard for these systems. This will come and for now, the USA utility scale inverter industry is to be regarded as experimental.

When connecting strings to the inverter, do not mix and match solar module types and connect them to a single inverter, this is a recipe for disaster. If you desire to install the different technology types on the same site, then each solar module technology must have its own inverter system dedicated to it.

Module mismatch in strings is also an issue for the inverter MPPT. Do not connect different wattage modules together in the same string as the higher wattage modules will try and push their current through the lower wattage ones. This will waste power. Each individual solar string needs to have identical wattage modules in it.

Foreign manufactured inverter systems may arrive with the inverter set for a different voltage or frequency of AC electricity. Always check that the inverter system matches the grid system that you are connecting into.

Switchgear

Switchgear is widely used in solar photovoltaic installations. A complete line of switchgear has been developed for this application. Specifically, combiner and re-combiner boxes are now listed amongst most manufacturers product lines.

Older switchgear was not constructed with solar power generation in mind and may only be rated for electricity flow in one direction. If so, they will have markings on the electrical terminals that say either "load" or "line".

Solar photovoltaic systems feature an AC bi-directional power flow into the grid that reverses at least twice per day. During the day time the solar power system will generate AC energy into the grid and during the night, the power flow will reverse and it will consume a small amount of power from the grid. On cloudy days this reversal will occur frequently during the day time on net zero systems.

As such, all switchgear in the AC circuit on a grid connected solar power system will need to be rated for bi-directional current flow. Due to their bi-directional operation, these electrical systems have no "load" or "line" markings on the electrical terminals.

Distribution

Distribution is a feature of the larger commercial installations and the utility installations. Basically distribution covers the medium voltage application which is normally at 24 kV in the USA.

If you are working at this level, it is extremely important that you are competent with working with this type of equipment. The voltages at this level are very unforgiving and death can easily result from misguided activities with this type of equipment. People die every year by electrocution on these systems, it is an occupational hazard. OSHA standard 1910.269 for electric power generation, transmission and distribution details the safety requirements for this type of equipment for the utilities. Work safe and be safe.

Solar photovoltaic systems feature a bi-directional current flow that reverses twice per day. During the day time the solar power system will generate energy into the grid and during the night, the current will reverse and it will consume a small amount of power from the grid. As such, all distribution equipment in a solar photovoltaic power system will need to be rated for bi-directional current flow.

Distribution generally covers the following items

- Transformers

- Distribution poles
- Medium voltage switchgear
- Relays
- Harmonics

Transformers

Dry type transformers are generally used in Solar photovoltaics and have very high efficiencies of over 99% conversion. Some people are still using oil filled transformers and the operation and maintenance costs are higher with these due to needing more attention and care.

Distribution Poles

Many installations use distribution poles to connect into the point of interconnection (POI) at the grid. It is a cheaper way of distribution when long distances are involved. For short distances of less than about 1 mile there really isn't any difference between using distribution poles or going underground, if the underground route is easy to put ducts and conduits into.

Medium Voltage Switchgear

Medium voltage switchgear is generally of the type:

- Air insulated
- Gas insulated

- Vacuum insulated

The choice of which type to use is generally driven by costs. The higher power switchgear is generally vacuum insulated.

Protective Relays

Protective relays control the tripping mechanisms on the switchgear. They are set up according to the desired fault levels that the designer has calculated for each switch on the system.

Harmonics

Inverter systems cause harmonics. Normally, total harmonic distortion (THD) should below 3% and is quite acceptable to feed power into the grid at these low levels.

Power Purchase Agreements (PPA)

Many solar installations are built using power purchase agreements (PPA) that are used to finance the project. A PPA is quite a simple idea. A loan to finance the project is taken out by the company that builds it. It is built on land that is owned by the customer. Once in operation and producing energy, that energy is sold to the customer who has agreed to purchase all energy produced by the system over twenty years. The sales of energy during the twenty years are used to pay for the operation and maintenance costs of the system and the original loan interest and principal.

After about ten years the loan is repayed in full and all future sales of energy produce profit at that point after routine costs. After the twenty years is up, the customer can either choose to enter into another PPA for energy supplied, have the system removed from their property or purchase the system.

The PPA is a great concept and when a system performs as planned or even better, they are a great way of building solar photovoltaic systems. Due diligence at the planning stage ensures this.

Site Licensing Agreements (SLA)

Site licensing agreements (SLA) are the partner to the power purchase agreement (PPA). If you are building on a piece of property that is owned by someone else, you will need one of these. It is basically a contract for land use for the duration of the project.

It is important that the SLA is well written and incorporates the following:

- Term of land use
- Unrestricted access for the duration of the project
- No impacts to the solar photovoltaic power generation system from future surrounding developments (generally shading & dirt problems)
- Energy costs
- Guaranteed power purchasing by the customer
- Customer maintains the point of AC interconnection in service at all times
- Customer pays compensation if the point of interconnection is not available for power production.
- End of lease needs to be well defined for the options to enter into another PPA, purchase the system, or have the system removed.

Health and Safety

Health and safety around solar photovoltaic systems is very important. In the DC circuit you will find up to 600 volts in residential and commercial systems. In utility systems you may find systems operating at close to 1,000 volts DC!

DC is very unforgiving and if you short the circuit, you may well start a fire. Most firefighting departments will not fight a solar photovoltaic fire until after sunset, due to these high voltages that are present on the system.

The DC circuit will keep producing power while it is exposed to sunlight and it is this attribute that makes the solar photovoltaic circuit a particular risk to personnel. You should be aware that the string fuses in the DC circuit will not blow under a short condition after the combiner box, due to the solar DC circuit being current limited. The fuses are there to protect against reverse current flow into the string circuit that could damage the solar modules.

You will commonly find yourself working on live systems and you should be taking the following precautions:

The top hazards in solar photovoltaics are:

- Falls.

Complete Solar Photovoltaics © Steven Magee

- Use of undersized and incorrect electrical equipment.
- Shocks and electrocution.
- Fire and arc flash.
- Excessive electromagnetic interference (EMI) exposure.

Regarding electromagnetic interference (EMI) exposure, this can cause Radio Wave Sickness (RWS) or Electromagnetic Hypersensitivity (EHS). The symptoms to look for are:

- Forgetfulness.
- Irregular heartbeats.
- Fatigue.
- Intoxication symptoms.
- Aggression.
- Anxiety.
- Seemingly random aches and pains.
- Arthritis symptoms, particularly in knees.
- Intestinal pains.
- Diarrhea.
- Headaches.
- Increased sexual desire.

If you start to show these symptoms, you should remove yourself from the solar field. You should attempt to identify the equipment that is causing the human health problems and establish if it is faulty. As equipment ages, the electromagnetic interference may increase from it. A Trifield meter and an AM radio that is tuned to static can be useful in detecting these problems.

You should be wary of using metal buildings and containers to house equipment in. These can cause very high electromagnetic interference (EMI) environments to occur and may make the personnel ill. Instead, mount all solar equipment outdoors using concrete pads and shade canopies. Try and use natural materials to construct the shade canopies. Metal roofing materials should be avoided as they act like reflectors to EMI and may increase the magnitude of it.

Do not cluster inverters together in groups, it is better to spread them out over the site on large installations. Leave at least three feet of clearance around each piece of electrical equipment for maintenance purposes.

Residential solar equipment should be mounted away from the human environment. It is preferable for electrical equipment to be mounted to the garage rather than the home. It is wise to keep the inverter out of the human environment and most definitely do not mount it on bedroom walls. If there is no suitable wall to mount it onto, then consider using micro-inverters or AC solar modules.

For roof mounted solar modules you should mount them near to the apex on a pitched roof to keep the EMI fields low in the human environment below. EMI decays with distance. For flat roofs, consider mounting them with a few feet between the lowest edge of the module and the roof surface.

Cleaning solar modules should not be done on a system that is producing power. The EMI effects and the risk of electrocution is at its greatest during cleaning operations with water. Instead, clean systems during the night time.

The diagrams on the following pages show the EMI fields being emitted by the inverter system.

INVERTER SYSTEMS AND HUMAN HEALTH

EXTENDED EXPOSURE TO INVERTER SYSTEMS MAY BE HARMFUL TO HUMAN HEALTH.

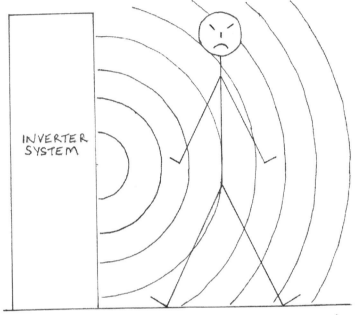

MANY INVERTERS CREATE ELECTROMAGNETIC FIELDS AROUND THEM. YOU SHOULD BE WARY ABOUT ENTERING THESE FIELDS.

© COPYRIGHT STEVEN MAGEE

Complete Solar Photovoltaics © Steven Magee

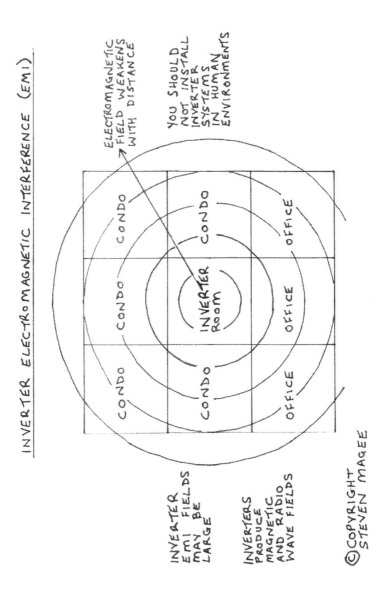

If you have overhead power lines, then these should be kept clear of tree growth. Arrange tree trimming in the fall, after the tree growth period ends. If growth is rapid, then you may have to trim more often. Trees growing into your power lines may lead to failures on the power system and fires.

Tree trimming is shown on the next page.

Tree Trimming

Specialized equipment is needed to trim the trees around aerial power lines. Test the lines for EMI before trimming and turn the power lines off if they are emitting significant EMI. Fix the EMI problems at the same time as trimming the trees.

You should have the following safety equipment:

- Insulated gloves.
- Insulated shoes.
- Insulated mats.
- Insulated tools.
- Hard hats.
- Eye protection (DC fuses can explode).
- Fall protection when working above 6 feet.
- Arc flash protective clothing.
- Reflective vests.
- Powder (ABC) fire extinguishers.
- First aid kit with burn treatments.
- Oscillscope with a Fast Fourier Transform (FFT) function.
- Trifield meter.
- AM Radio.

Always tie off your ladders. Make sure that you know the layout of the roof and where weak spots such as roof skylights are. Do not step on roof skylights and roof vents and consider roping them off. Use cones and reflective tapes to mark out roof hazards that are to be avoided.

OSHA has a solar safety wesite at:

Complete Solar Photovoltaics © Steven Magee

www.osha.gov/dep/greenjobs/solar.html

The following picture shows the correct site dress code.

Safety Clothing and Tools

Safety clothing and the correct tools are needed around solar photovoltaic systems. These people were trimming trees around power lines and should probably be wearing long sleeve arc flash shirts.

You should not spend time directly under power lines, as this will put you into the plasma field. Plasma is the fourth state of matter and under the power lines is an invisible flow of electrons from the lines into the ground below through capacitive coupling. It is the reason why florescent tubes will light there.

During my research into power lines producing AM radio frequencies, I noticed the reflection effect. The cell phone tower microwave signals seem to be interacting with the power lines and may be producing pockets of AM radio frequencies that can be picked up on a standard AM radio tuned to static. If you were in one of these pockets for an extended time period, you may develop Radio Wave Sickness (RWS) or Electromagnetic Hypersensitivity (EHS).

When working around power generation, distribution and transmission systems it is wise to establish the presence of the various fields and to exercise caution around them. Stay out of strong electromagnetic interference (EMI) fields as they may eventually make you sick.

The following pages demonstrate the various effects of power lines.

Complete Solar Photovoltaics © Steven Magee

Power lines and poles can have many types of large fields around them. "Dirty electricity" effects may cause extensive radio wave fields.

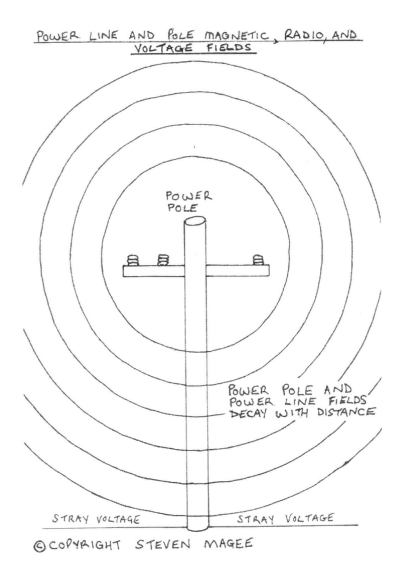

Complete Solar Photovoltaics © Steven Magee

Power Pole Metal Work

Induction effects in power pole metal work may cause sparks to jump between it which will cause radio wave emissions to occur. Defective insulators do the same.

Complete Solar Photovoltaics © Steven Magee

Power lines and poles can emit plasma and ions. The high voltage causes the electrostatic attraction effect. Power lines and poles have fields that extend out from the area that set backs should be applied to, to protect human health.

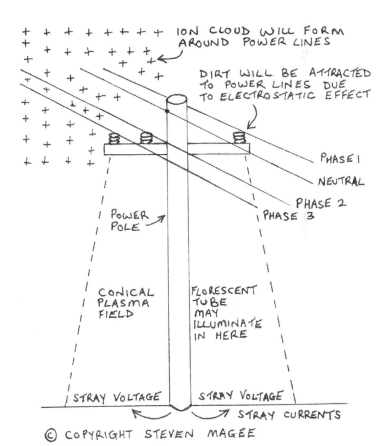

Complete Solar Photovoltaics © Steven Magee

Power Line and Pole Solar Interference

The power poles and lines can interfere with the solar radiation transmission when in front of the Sun.

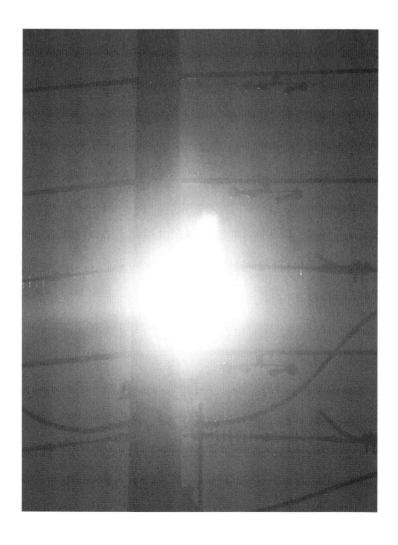

Complete Solar Photovoltaics © Steven Magee

Power Line Reflections

Power lines may cause radio and microwave reflections and interference effects to occur.

Complete Solar Photovoltaics © Steven Magee

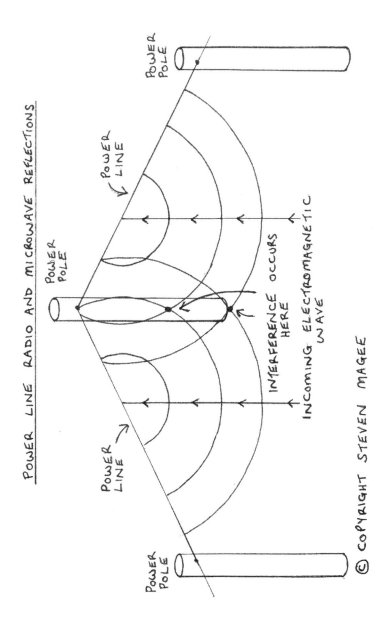

Dr. Phillip Stoddard, Professor of Biological Sciences at Florida International University, has done extensive research on power lines. He has found very significant health risks from their presence:

- The closer you live to a power line, the more likely you are to develop leukemia.
- Living in a magnetic field of 3.5 milli-gauss doubles the leukemia risk.
- Living within 0-50 meters of a power line doubles the risk of Alzhiemer's Disease and presents a 1.5 increased risk of developing senile dementia.
- Burying the power lines brings the magnetic fields closer.

A common problem on buried power lines is the corrosion of the concentric neutral. The concentric neutral is the wire that you see wrapped around the outside of the utility cable that comes down the power pole. If this corrodes, then the neutral starts to become high resistance and this will cause current to increase though the ground. Basically, corrosion of the concentric neutral will electrify the surrounding ground and is clearly a human health hazard. Faulty insulation on the live conductor that causes leakage currents will have a similar effect. You should be aware of these effects when working around electrical systems and take the appropriate precautions.

Complete Solar Photovoltaics © Steven Magee

In particular you should limit your time in these environments:

- Power lines.
- Power poles.
- Transformers.
- Substations.
- Switch yards.
- Power generation plants.

The areas in the vicinity of these may have large amounts of stray voltages and currents in the ground.

It is important when designing power systems that you use setbacks to prevent the electrical system from impacting both the health of the electrical workers and the people who may be around the system daily. You should consider fencing the setbacks to prevent people from inadvertently entering them.

Planning Requirements

There are many aspects to designing a solar photovoltaic system and you should be aware of each aspect and the costs that these bring. Some of the items on the list will be needed and others may not. You should consult with the authority having jurisdiction (AHJ) to verify the requirements:

- Rebates and subsidies for project.
- Soil analysis.
- Flood plain requirements.
- Storm Water Prevention Plan (SWPP).
- Electrical grounding test.
- Site boundary survey.
- Site slope and grading plan.
- Setbacks.
- Ground clearance for electrical equipment.
- Glare assessment.
- Wildlife survey.
- Environmental Impact Study (EIS).
- Archeological survey.
- Fire Permit.
- Lightning protection.
- Weather system.
- Security system.

Complete Solar Photovoltaics © Steven Magee

- System monitoring.
- Wind and snow loading.
- Seismic requirements.
- Temperature requirements.
- Site altitude corrections.
- Shading impacts.
- Electrical interconnection requirements.
- Foundation requirements.
- Fencing requirements.
- Distribution and transmission requirements.
- Heavy equipment requirements.
- Site Access.
- Equipment lead times.
- Construction staging areas.
- Construction storage requirements.
- Site security required.
- Site office.
- Site restrooms.
- Maintenance requirements.
- Planning department application process.
- Utility application process.

Planning a project can be extensive and some projects need years of planning and surveys before you can even consider starting construction on your designed system. You should establish at the very start of the design process the requirements for the site and the lead times and costs for each requirement. The authority having jurisdiction (AHJ) should be able to give you the requirements needed for building the project.

When submitting your plans for the permitting process, you should be aware that the permitting officer may not be aware of all of the codes and regulations for solar photovoltaic systems. As such, you should submit a complete plan package to them that clearly references the following:

- The current site conditions and how it will change.
- Calculations and formulae.
- Use the definitions in NEC 100 and NEC 690.2 to label your diagrams.
- Have large commercial and utility system plans stamped by a Professional Engineer (PE) who specializes in solar photovoltaic systems.
- List all code standards used and their versions.
- Plans can generally be neatly hand drawn or computer (CAD) drawn.
- For multiple inverter systems that are identical, one set of drawings of a single inverter system is generally sufficient.

Be aware of the time taken for plans to be approved and put contingency in for resubmitting them if changes are deemed necessary.

Bill Brookes has written the "Expedited Permit Process for PV Systems" and it can be obtained from:

www.solarabcs.org/permitting

Lightning protection is desirable on the systems and most inverter systems have AC and DC lightning arrestors built into them. You should try to select DC combiner boxes that have this feature too. Some people also install lightning rods to the north of the solar array. If you want to find out more about designing for lightning, obtain a copy of "NFPA 780: Standard for the Installation of Lightning Protection Systems."

It is important that you are competent in solar photovoltaic design. You should not be using software to design solar photovoltaic systems if you do not understand how to perform the calculations manually. Using software may get you into trouble if you cannot verify the results!

The current software tends to be tricky to use and it takes a lot of practice to become proficient in the various software offerings that are available. I personally prefer to design systems by manually performing the calculations as it is generally faster. The software can be useful for verifying your calculations.

The various types of software that are used to design systems can give results that can vary by up to 14% between products for annual performance predictions. They are all using their own algorithms to generate their performance predictions that vary widely between products. Annual production figures are rarely accurate due to the variances in the weather from year to year.

Unfortunately, it is common in the industry for annual system performance to be overstated in order to bring in sales and financing for the systems. Realistic efficiencies for solar power systems lie between 60% to 85% conversion from solar irradiance into AC grid power at the point of interconnection and it depends on location, components used, and the quality of design and construction.

Typical Meteorological Year (TMY) 3 data is normally used to predict the annual system performance. There are 1,020 USA locations and three types of data sets. Class 1 is the most complete, class 2 is less certain, and class 3 is incomplete data. The global horizontal value is what most solar photovoltaic designers are interested in and it comprises of direct beam radiation (>90%) and horizontal diffuse radiation (<10%).

Complete Solar Photovoltaics © Steven Magee

Residential Design

Residential design for the purposes of this book is systems up to 10 kWp DC at STC. Most readers of this book will be working with this size of system. Residential systems represent the starting point for our designs in this book and underpin the larger system designs. To understand solar design, this is your starting point.

Residential systems are relatively easy to design electrically and have very few solar strings, sometimes they may only have one string. Something to consider with a residential system is the size of the solar modules that you will use. Since most solar systems on a residential system will be mounted to the homeowners roof, it is wise to select a solar module that is easily handled when up on the roof. Sometimes bigger is not better when installing roof mounted systems.

A couple of points to be aware of in the NEC:

NEC 690.11 added DC arc fault circuit protection to solar power systems on or penetrating a building. Since this requirement was only added in 2011, it will be a few years before these products will appear in the market place. You will need to check with the authority having jurisdiction if this applies to your installation.

NEC 690.31(E) DC photovoltaic source and output circuits inside a building need to be run in a metal raceway. It needs to be at least 10" below the roof decking or sheathing. It must be marked as containing solar photovoltaic circuits along its length.

For this example we will assume that this will be a tiled, thirty degree slope, South facing roof mounted system. An assessment must be made to make sure that the roof is strong enough to mount the solar power system to. A structural engineer can perform this assessment.

Once confirmed that we can mount a solar photovoltaic system to the roof, you will need to identify where the roof supports or trusses run. You will want to attach your mountings to this location. A roofer can help you with removing the tiles and attaching your chosen mounting to the roof. Use roof sealant under your standoffs. Sometimes you will be able to drill the tile and refit it, other times you will want the roofer to flash the mounting with lead to waterproof the location in the absence of the tile being refitted. Chalk and grease pens are useful for marking out roof systems.

The system that we will design will be a 5 kWp DC system which will be typical for most homes.

The solar module specifications are as follows:

- Maximum Power at STC = 235 W
- Tolerance of Power = +10% / -5%
- Open circuit voltage at STC = 37V
- Maximum power voltage at STC = 30V

- Short Circuit Current at STC = 8.6A
- Maximum power current at STC = 7.84A
- Maximum system voltage = 600VDC
- Series fuse rating = 15A
- Power temperature coefficient = -0.485%/°C
- Voltage temperature coefficient = -0.351%/°C
- Current temperature coefficient = 0.053%/°C
- Normal operating cell temperature (NOCT) = 47.5 °C
- Size (L x W x H): 48 in x 36 in x 1 in

Our Inverter specifications are:

DC Specifications:-

- Continuous Power @ 240 VAC: 5150W
- Recommended Max PV (STC): 6200W
- MPPT Voltage Range: 200V - 550V
- Maximum Input Voltage: 600 VDC
- Strike Voltage: 235 VDC
- Maximum Input Current: 25A
- Maximum Input Short Circuit Current: 30A
- Fused Inputs: 4

AC Specifications:-

- Continuous Power @ 240 VAC: 4900W
- Voltage Range @ 240 VAC: 211-264VAC

- Frequency Range: 59.3-60.5Hz
- Continuous Current: 20.7A
- Output Current Protection: 30A
- Max Backfeed Current to PV: 0A
- Power Factor: Unity >0.99
- Total Harmonic Distortion: <3%
- Efficiency: 96%

General:-

- Enclosure: Rainproof, NEMA 3R
- Housing Material: Painted Aluminum
- Ambient Temperature Range: -25°C to +55°C
- Weight: 61.7 lbs, 28Kg
- Cooling: Convection & Fan Assist
- Wire Sizes: 12 to 6 AWG input & output connections
- Size (L x W x H): 28 4/5 in x 17 ¾ in x 8 ¼ in (732mm x 454mm x 210mm)
- Standards: UL1741/IEEE1547, IEEE1547.1, ANSI62.41.2, FCC Part 15B
- Warranty: 15 Years

Location Area Specifications are:

- Historic annual minimum temperature: -10°C
- Historic annual maximum temperature: +40°C

- Historic annual maximum wind speed: 80 MPH
- Historic annual snow fall depth: None
- Historic annual hail size: None
- Latitude: 30
- Reflections: None

Of note is how the solar module ratings vary with temperature. Voltage is affected the most with current almost unaffected.

This solar module has a normal operating cell temperature (NOCT) of 47.5 degrees Celsius. This is 22.5 degrees above the STC rating. This gives:

- NOCT Power: 210 W
- NOCT Voltage: 34.1 V
- NOCT Current: 8.7 A

As we can see, our solar module looks quite different at a higher temperature than STC. This is an important concept to grasp in solar photovoltaic design.

It is not just the solar modules that are affected by temperature. Our equipment such as terminals, cables, fuses, and so on are affected also. As such, we will use the following values for our system design de-ratings for this area:

- 80°C Attic temperature and solar module temperatures
- 60°C Enclosure temperature
- 40°C Underground temperature

Now onto the design of our system:

To get a 5 kWp DC system at STC, we need to divide 5 kW by STC solar module power:

5,000W / 235W = 22 modules minimum

We need to put our strings together for a voltage below 600 VDC. This requires an adjustment to the open circuit voltage to allowing for a minimum temperature for this system of -10 Celsius

Temperature difference = STC − Lowest temperature

=25°C - (-10°C) = 35°C

Open circuit voltage adjustment = -35°C x (-0.351%)

= 12.285% adjustment

Solar module maximum STC voltage = V_{oc} = 37V

Adjusted solar module maximum voltage = 37V * 112.285%

= 41.55V

We call a group of solar modules electrically connected together in series a "string". We do this in order to increase the DC voltage to a level that the inverter can use. So our maximum number of solar modules in a string is:

= System voltage / adjusted solar module maximum voltage

= 600V / 41.55V

= 14 solar modules maximum in strings

Our solar module is 235 Watts, so this gives a string wattage at STC of:

String wattage = Number of solar modules x module watts

= 14 x 235W = 3,290W

String voltage = Number of solar modules x adjusted solar module maximum voltage

= 14 x 41.55V = 581.7V

Number of strings in system = System watts / string watts

= 5,000W / 3,290W = 1.5

We need to round up to make our strings equal, so we would go to two strings at this point:

5,000W / 2 = 2,500W

Each string needs to be at least 2,500W. We divide this figure by our module wattage to get our string number:

2,500W / 235W = 10.63 modules

We round up to 11 modules in each string to give:

String watts = 11 modules x 235W = 2585W

Two strings = 2 Strings x 2585W = 5170 Watts

String maximum voltage = adjusted maximum voltage x number of modules in string

= 41.55V * 11 = 457V

Complete Solar Photovoltaics © Steven Magee

Solar modules generate their maximum voltage at their lowest temperature. Conversely, they generate their minimum voltage at their highest temperature.

Assuming that the maximum temperature that the solar module will operate at is 40°C higher than ambient temperature:

Solar module adjusted minimum voltage = STC MPPT voltage - ((Annual maximum temperature + 40°C increased module temperature adjustment - 25°C STC temperature) x voltage temperature adjustment percentage x STC MPPT voltage)

= 30V + ((40°C + 40°C - 25°C) x (-0.351%) x 30V))

= 24.2V

String minimum voltage = number of strings x solar module adjusted minimum voltage

= 11 x 24.2V

=266.2V

Looking at our minimum and maximum values and comparing it to the inverter DC voltage specifications, both values will work well with this inverter.

Complete Solar Photovoltaics © Steven Magee

266.2V Minimum string voltage > 235V Inverter strike DC voltage

457V Maximum string voltage < 600V Maximum input DC voltage

Next we look at our maximum current for the string. From the data sheet for the solar module we see that the solar module short circuit current at STC is 8.6A with an adjustment for current temperature coefficient of 0.053%/°C. Now let's see how much current we should expect from the system at our maximum ambient temperature of 40°C.

Adjusted maximum solar module current output = Short circuit current at STC + ((annual maximum temperature + 40°C increased module temperature adjustment - 25°C STC temperature) x (current temperature coefficient x STC short circuit current)

= 8.6A + ((40°C + 40°C − 25°C) x 0.053% x 8.6A

= 8.85A

As you can see, there is very little effect on current by the increased temperature of the module.

We now calculate the maximum continuous circuit current in NEC690.8(A) which gives:

Maximum continuous circuit current = Short circuit current x 125%

= 8.6A x 125%

= 10.75A

The National Electric Code requires the fuses and cables to be rated at least 125% higher than the maximum continuous circuit current. Per NEC690.8(B) this gives:

Overcurrent device ratings = Maximum continuous circuit current x 125%

= 10.75A x 125%

= 13.44A

The National Electric Code includes this 25% factor for fuse and cable de-rating. We will use the 15A maximum fuse as recommended by the manufacturer.

String fuses =15A

This is higher than needed and incorporates sufficient de-rating for the fuse for this application.

The cable amperage size should be the same or larger than the fuse size. Since our system is mounted to the roof and our cables will pass through the attic in metal conduit, we will use NEC table 310.16 to obtain our cable size. Since the attic is a hot location, we will de-rate the cable for a maximum temperature of 80°C.

Cable current =>15A

80°C de-rating = 0.41

De-rated cable size = fuse size / 80°C de-rating

= 15A / 0.41

= 36.6A

Looking at NEC table 310.16 we see that 10 AWG cable exceeds this at 40 amps.

String cable size = 90°C 10 AWG

Due to the short run to the inverter we will simply run each string circuit in its own conduit. This avoids the need to de-rate further. If you were to put both strings in one conduit you would have to apply the 4 conductor derating in table 310.15(B)(2)(a) of 0.8. This would increase the cable sizes to the next size up. It is preferable in the DC circuit to limit the number of connections for reliability purposes and by running two conduits, we avoid installing a transition box.

Finally, our electrical ground conductors are sized. There are two ground conductors used, one for the system and one for the equipment.

The system ground is achieved by connecting one conductor of our two wire DC system to ground. In this particular case, this is achieved internally inside the inverter using the negative conductor.

The equipment ground is to protect all exposed non current carrying metal parts in the system, such as module frames. When grounding the module frames, use the recommended grounding lug that the manufacturer specifies in their installation manual. NEC 690.45 specifies the size of the equipment ground cable to be obtained from table 250.122.

From NEC table 250.122 we obtain our equipment ground cable size and we see that 14 AWG copper cable is needed for a 15A fused circuit.

Equipment ground cable size = 14 AWG

For each individual string we will use a 14 AWG cable to our ground location at the inverter.

The support racking for the solar modules will be grounded in accordance with the manufacturers instructions and local codes.

Complete Solar Photovoltaics © Steven Magee

The voltage drops are less than 3% in this application since there is only 100 feet of cable between the solar modules and the inverter.

Since the distance from the solar modules to the inverter is so short, we will use the inverter inputs to combine the two solar string circuits. We will install a DC disconnect switch at the nearest accessible location to the solar panels.

Now onto the AC circuit:

Our connection point to the grid will be at the utility meter. This eliminates any need to verify if the fuse board is capable of being back fed by the solar photovoltaic system. Our inverter will be mounted next to the meter with a AC fused disconnect switch between them. The fuses will be sized according to the inverter output fusing specification of 30A. Our cable needs to be sized to at least 30 amps also.

Cable size = fuse size x 60°C cable de-rating

= 30A / 0.71

= 43A

Looking at NEC table 310.16, we see that 90°C 8 AWG copper cable will meet this need at 55A.

Equipment ground cable is obtained from NEC table 250.122 and is 10 AWG copper for a 30A breaker.

We do not need to install a solar power grounding system as we will connect into the existing household grounding electrode system. If we were to install a grounding electrode system, it would use a minimum 8AWG copper grounding electrode conductor per NEC 250.166. The DC grounding system is bonded to the AC grounding system.

The annual maximum wind speed is 80 MPH and a solar module mounting system will be used that exceeds this requirement.

So here are our system specifications:

- Inverter: 5 kW, single phase, 240 VAC, 600 VDC
- Inverter 240 VAC cabling: 90°C 8 AWG copper
- Inverter AC ground cable: 10 AWG copper
- Solar string cabling: 90°C 10 AWG copper
- Solar string equipment ground cable: 14 AWG copper
- Solar string fuses: 15A
- Number of solar string fuses: 2
- Number of solar strings: 2
- Number of solar modules: 22

Complete Solar Photovoltaics © Steven Magee

Have your solar photovoltaic design checked by your local solar photovoltaic licensed electrician before submitting the plans for approval to you local authority.

The system can now be drawn and is shown on the following page.

Complete Solar Photovoltaics © Steven Magee

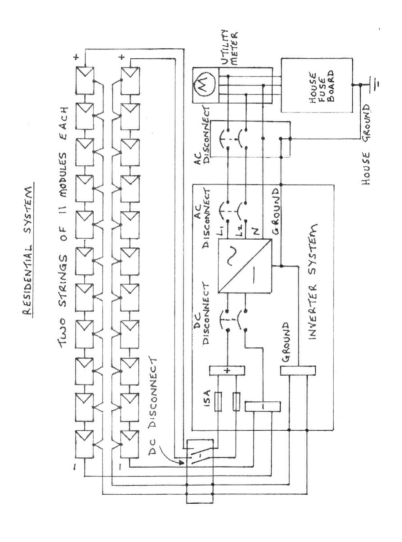

Complete Solar Photovoltaics © Steven Magee

Typical Home Inverter System

This picture shows a typical home inverter system.

Commercial Design

Commercial design for the purposes of this book are systems from 10 kWp DC to 1 MWp DC at STC.

New ideas in this section are

- MPPT tracking performance
- Combiner boxes

MPPT Tracking Performance

Maximum power point tracking (MPPT) is at its most efficient on a single solar module string. As you increase your system size and add more strings to it, it can reduce the performance of MPPT. The MPPT tracking will only be as good as the worst performing strings connected to it. As such it is important to limit the number of strings connected to an inverter. I would suggest that no more than one hundred strings be connected to an inverter for this reason.

Combiner boxes

These are needed to connect our strings together so that they can be combined into a larger cable. They generally come in 6, 12, 18 and 24 string capacities. More advanced models have the capacity to monitor each string connected to it using a computer network and the associated software. Each string will have a fuse or circuit breaker for it. UL 1741 applies to combiner boxes and look for products with this marking.

A 12 input combiner box wiring diagram is shown on the next page. The following page shows that the conduits start to get very large on systems this size.

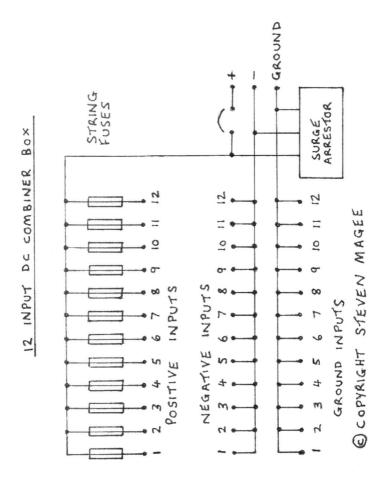

Complete Solar Photovoltaics © Steven Magee

Conduits

Conduits, trunking and junction boxes may start to get large on these installations.

250 kW System Design

Our commercial system will be a single 250 kW inverter design that will connect into the distribution system of the commercial facility.

We will use the same solar module as before and this time we will increase the number of modules in the string to the maximum voltage that the inverter can accept.

This will be a ground mounted, 30 degree fixed tilt system that connects into the commercial facility electrical system which is a 480 volt, three phase grounded system. We have verified that the existing facility switchgear is capable of being back fed by the solar photovoltaic system and that the bus system can handle the current passing through it.

Inverter Specifications

Our inverter specifications are :

Input Parameters:-

- PV array configuration: Negative ground
- MPPT Voltage Range: 320V - 600VDC
- Maximum Input Voltage: 600 VDC
- Maximum Input Current: 814A

Output Parameters:-

- Maximum Continuous Output Power: 250kW (250kVA)
- Voltage Range @ 480 VAC: 422 – 528 VAC
- Nominal Voltage: 480 VAC
- AC Voltage Range: -12%/+10%
- Frequency Range: 59.3-60.5Hz
- Nominal Output Frequency: 60 Hz
- Number of Phases: 3
- Maximum Output Current Per Phase: 301A
- Power Factor at Full Load: >0.99
- Total Harmonic Distortion: <3%
- Efficiency: 96%

Temperature:-

- Operating Ambient Temperature Range (Full Power): -20°C to +55°C
- Storage Temperature Range: -30°C to +70°C
- Cooling: Forced Air

Noise:-

- Noise Level: <65dB(A)

Combiner:-

- Number of Inputs and Fuse Rating: 10 (160ADC)
- Enclosure Rating: NEMA 3R, IP 44
- Enclosure Finish: RAL-7032
- Inverter Cabinet Dimensions (Height x Width x Depth): 92.6" x 117.7" x 43.3" (235.2mm x 298.96mm x 109.98mm)
- Inverter Cabinet Weight: 4500 lbs (2,041 kg)

Testing Certifications:-

- Standards: UL1741, CSA 107.1-01, IEEE1547, IEEE C62.41.2, IEEE C62.45, IEEE C37.90.1, IEEE C37.90.2
- UBC Zone 4 Seismic Rating

Warranty:-

- Five Years

Monitoring:-

- Third Party Compatible

This inverter has an internal transformer that enables us to connect the inverter directly into the electrical system.

To get a 250 kWp DC system at STC, we need to divide 250 kW by the solar module STC power:

250,000W / 235W = 1,064 solar modules minimum

We need to put our strings together for a voltage below 600 VDC. This requires an adjustment to the open circuit voltage to allow for a minimum temperature for this system of -10 Celsius.

Temperature difference = STC – lowest ambient temperature

=25°C - (-10°C) = 35°C

Adjustment: -35°C x (-0.351%) = 12.285% adjustment

Solar module maximum STC voltage = 37V = V_{oc}

Adjusted solar module maximum voltage = 37V x 112.285%

= 41.55V

Maximum number of solar modules in string = inverter maximum DC voltage / adjusted solar module maximum voltage

= 600V / 41.55V

= 14 solar modules in strings

Our solar module is 235 Watts, so this gives a string wattage at STC of:

String wattage = number of solar modules x module watts

= 14 x 235W = 3,290W

String voltage = number of solar modules x adjusted voltage

= 14 x 41.55V = 581.7V

Number of strings in system = system watts / string watts

= 250,000W/ 3,290W = 76 strings

Total DC power = 76 Strings x 3290 W = 250,040 Watts

String maximum voltage = adjusted maximum voltage x number of modules in string

= 41.55V x 14 = 581.7V

Solar modules generate their maximum voltage at their lowest temperature. Conversely, they generate their minimum voltage at their maximum temperature.

Assuming that the maximum temperature that the solar module will operate at is 30°C higher than ambient temperature:

Solar module adjusted minimum voltage = STC MPPT voltage - ((Annual maximum temperature + 30°C increased module temperature adjustment - 25°C STC temperature) x voltage temperature adjustment percentage x STC MPPT voltage))

= 30V+((40°C + 30°C - 25°C) x (-0.351%) x 30V)

=25.3V

String minimum voltage = number of strings x solar module adjusted minimum voltage

= 14 x 25.3V

= 354.2V

Looking at our minimum and maximum values and comparing the to the inverter DC voltage specifications, both values will work well with this inverter.

354.2V Minimum string voltage > 320V inverter minimum DC voltage

581.7V Maximum string voltage < 600V inverter maximum input DC voltage

We now adjust for maximum continuous current using NEC 690.8(A) which gives:

Continuous current = maximum solar module short circuit current output x 125%

= 8.6A x 125%

= 10.75A

The National Electric Code 690.8(B) requires the fuses and cables to be rated at least 125% higher than our normal circuit current. This gives:

NEC circuit sizing and current = 10.75A x 125%

= 13.44A

This is the factor for fuse de-rating. If the manufacturer specifies fuse de-ratings for certain temperatures, then this value should be calculated and the larger value of the two calculations should be used.

Using the solar module manufacturers 15A maximum fuse is what we will select for this application.

String fuses =15A

Complete Solar Photovoltaics © Steven Magee

The cable size should be the same or larger than the fuse size, as this is what the fuse protects. Since our system is ground mounted and our cables will pass through raceways we will use NEC table 310.16 to obtain our cable size. The conduits are in a high ambient location and we will de-rate the cables for a maximum ambient temperature of 60°C.

Cable current => 15A fuse size

60°C de-rating = 0.71

De-rated cable size = Cable current / 60°C de-rating

= 15A / 0.71

= 21.12A

Looking at NEC table 310.16 we see that 90°C, 14 AWG copper cable exceeds this at 25 amps.

String cable size = 90°C 14 AWG copper cable

Finally, our electrical ground conductors are sized. There are two ground conductors used, one for the system and one for the equipment.

The system ground is achieved by connecting one conductor of our two wire DC system to ground. In this particular case, this is achieved internally inside the inverter. The negative cable is grounded when the system is in operation.

The equipment ground is to protect all exposed non current carrying metal parts in the system, such as module frames. When grounding the solar module frames, use the recommended grounding lug that the manufacturer specifies in their installation manual. NEC 690.45 specifies the size of the ground to be obtained from table 250.122.

NEC table 250.122 indicates a 14 AWG copper cable for a 15A fuse, so this is what we will use.

Equipment ground cable size = 14 AWG copper cable

We will use the ground clamps that the manufacturer specifies for the solar modules and the ground will be a continuous unbroken cable.

The mounting structure will be grounded in accordance with the manufacturers instructions and local codes.

We will now combine the solar strings into a larger cable using the combiner boxes. The outputs of the combiners will pass underground to the inverter.

Continuous current output of 10 string combiner:

10 Combiner = 10 strings x short circuit current x 125%

= 10 x 8.6A x 125%

= 107.5A

Our fuse and cable size will be a further multiple of 125% per NEC 690.8(B)

Fuse size = continuous current output x 125%

= 107.5A x 125%

= 135A

From NEC 240.6(A) we see that the next standard fuse size is 150A and we will use this.

We match our cable size with the fuse size and de-rate it for an underground temperature of 40°C:

10 combiner de-rated cable size = fuse size / 40°C de-rating

= 150A / 0.91

= 165A

Looking at NEC table 310.16, we see that a copper cable size of 90°C, AWG 1/0 has a rating of 170A and is suitable for this application.

10 combiner box cable size = 90°C 1/0 AWG copper

Now for the ground conductor per NEC table 250.122.

From NEC table 250.122 we see that a 6 AWG copper cable is used with 150 amp fuse and this is what we will use.

Now let's do the same for the 9 combiner box:

9 Combiner continuous current = 9 strings x short circuit current x 125%

= 9 x 8.6A x 125%

= 96.75A

9 combiner fuse size = 96.75A x 125%

= 121A

NEC 240.6(A) next standard fuse size is 125A and this is what we will use.

9 combiner de-rated cable = combiner fuse / 40°C derating

= 125A / 0.91

= 138A

Looking at NEC table 310.16, we see that a copper cable size of 90°C 1 AWG has a rating of 150A and is suitable for this application.

9 combiner box cable size = 90°C 1 AWG copper

Now for the system ground conductor. Table 250.122 specifies #6 copper conductor for a 125A fuse.

Now onto the AC circuit:

Our connection point to the grid will be at the utility meter distribution center. We will use a spare distribution switch to feed our inverter which will be located 100 feet away. The fuse size for this switch will be:

Inverter AC Fuses = Maximum Phase Current x 125%

= 301A x 125%

= 376A

Using NEC 240.6(A) we see that our next standard size fuse is 400A and this is what we will use. If the fuse manufacturer supplies fuse temperature de-ratings, calculate this value and use the largest of the two values for the fuse size.

Our cable size will be 400A with a de-rating for 40°C for being underground.

Cable size = fuse size / cable temperature de-rating

= 400A / 0.91

= 440 amps

From NEC table 310.16, we see that 90°C 600 kcmil copper cable meets this requirement at 475 amps.

Now for the system ground conductor. Per NEC table 250.122 we see that a 3 AWG copper cable is specified for a 400 amp fuse.

The inverter will be grounded in accordance with the manufacturers instructions and local codes.

Looking at our annual maximum windspeed of 80 MPH, a solar module mounting system will be used that exceeds this requirement. The mounting system will be grounded in accordance with the manufacturers instructions and local codes.

The design that we are looking for is to be a long to run the length of the commercial premises on the south side. We are located at 30 degrees latitude so the solar modules will be tilted at 30 degrees, facing south.

4 divides well into 76 strings and this will give a design of:

76 strings = 4 rows x 19 strings

Looking into sizing this, our solar modules will be 3 feet wide. We will leave a 1 inch gap between each module for ventilation and expansion.

Row length = (number of strings in row x number of solar modules in a string x (solar module width + 1 / 12)) - 1/12

= (19 x 14 x (3 + 1/ 12)) - 1/12

= 820.17 feet long

We will place our inverter to the north of the modules to prevent shading from it and in the center of the row to minimize long cable runs. To the East on each row we will have ten strings and to the West we will have nine strings. Our combiner boxes will be 4 combiners of 10 strings and 4 combiners of 9 strings.

We need to make sure that our shading from each row is not affecting the row behind. To do this we use the equation:

Row distance = sin (sun elevation + solar module tilt) x solar module length / sin (sun elevation)

At this point you will want to reference a sun path chart. This is shown on the next page and was obtained from http://solardat.uoregon.edu/SunChartProgram.html at the University of Oregon Solar Radiation Monitoring Laboratory.

Looking at the solar chart on the next page we can see that in wintertime that the Sun will be above 5 degrees about half an hour after sun rise and half an hour before sunset. If space was tight, you could put the rows closer together, but you do want your system completely unshaded about an hour after sunrise through to an hour before sunset. In this location that equates to about 12 degrees.

Complete Solar Photovoltaics © Steven Magee

Solar Path Chart

Source: http://solardat.uoregon.edu/SunChartProgram.html at the University of Oregon Solar Radiation Monitoring Laboratory.

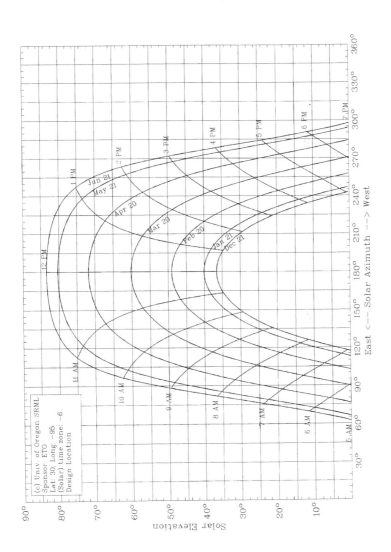

We want our solar modules to function without shading after the sun is above 5 degrees of the horizon. Our solar module is 4 feet long and is mounted at 30 degrees tilt. This gives:

Row Distance = sin (5 + 30) x 4' / sin 5

= 26.33 feet

This measurement is from the front of one solar module to the front of the next.

We have three spaces between our four rows of solar modules, so we will cover

Total separation distance = number of spaces between rows x row separation distance

= 3 x 26.33 feet

= 79 feet

On this system we will have some long cable runs, some up to 432 feet from the inverter location and we will need to check on the voltage drop for the solar modules that are furthest out from the inverter.

Volt drop = (2 x length x resistance per 1,000 feet x current) / 1,000

Complete Solar Photovoltaics © Steven Magee

Volt drop = (2 x L x R x I) / 1,000

Note that the cable length L is for one direction only. The equation multiplies L by 2 for the round trip.

Since our longest run is 432 feet on the DC cabling, we will calculate for this. Chapter 9 of the NEC has cable tables in it. Table 8 is for DC conductor properties. Looking at table 8, we see that #14 copper, 7 stranded cable has a resistance of 3.26 ohm per thousand feet. We will use the solar module Isc x 125% value for continuous current. The longest cable run on the modules is 432 feet.

DC volt drop = (2 x 432 x 3.26 x 10.75A) / 1,000

DC volt drop = 30V

If we move up to a #12 cable we will reduce the volt drop on the longest cable runs to a lower value. Looking at table 8, we see that #12 copper, 7 stranded cable has a resistance of 2.05 ohm per thousand feet.

DC volt drop = (2 x 432 x 2.05 x 10.75A) / 1,000

DC volt drop = 19V

If we move up to a #10 cable we will reduce the volt drop on the longest cable runs to a lower value. Looking at table 8, we see that #10 copper, 7 stranded cable has a resistance of 1.29 ohm per thousand feet.

DC volt drop = (2 x 432 x 1.29 x 10.75A) / 1,000

DC Volt drop = 12V

Given that this system has a mixture of long, medium and short cable runs, we will use the following cable sizes in the DC circuit:

- Up to 150 feet = 14 AWG 90 C copper
- Up to 300 feet = 12 AWG 90 C copper
- Over 300 feet = 10 AWG 90 C copper

This will assist the inverter MPPT tracking system at keeping the MPPT voltage correct throughout the DC circuit.

So here are our system specifications:

- Inverter: 250 kW, 3 phase, 480 VAC, 600 VDC with internal transformer
- Inverter 480VAC cabling: 600 kcmil 90°C copper
- Inverter equipment ground cabling: #3 copper
- Inverter DC cabling for 10 input combiner: 90°C 1/0 AWG copper

- Inverter equipment ground cabling for 10 input combiner: 6 AWG copper
- Number of 10 input combiner boxes: 4
- Inverter DC cabling for 9 input combiner: 90°C 1 AWG copper
- Inverter equipment ground cabling for 9 input combiner: 6 AWG copper
- Number of 9 input combiner boxes: 4
- Solar string cabling: 90°C 14-10 AWG copper
- Solar equipment ground cabling: #14 copper
- Solar string fuses: 15A
- Number of solar string fuses: 76
- Number of solar strings: 76
- Number of solar modules: 1,064
- Row Spacing: 26.7 feet

When the utility provides the interconnection details you will be able to calculate the fault currents in the design. With these values the correct switchgear can be selected, ensuring that it is rated for back feed.

Once the design is completed, it will need to be stamped by a licensed Professional Engineer to verify the design and the calculations.

The system can now be drawn and is shown in the following pages.

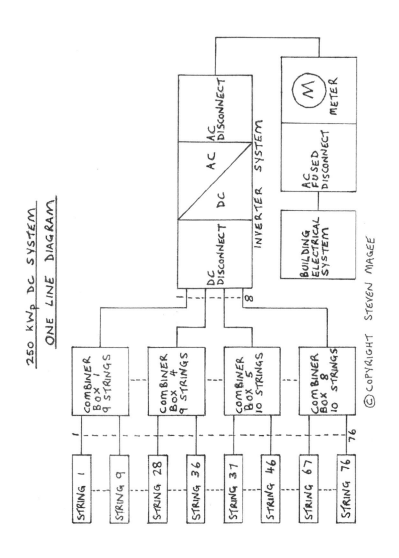

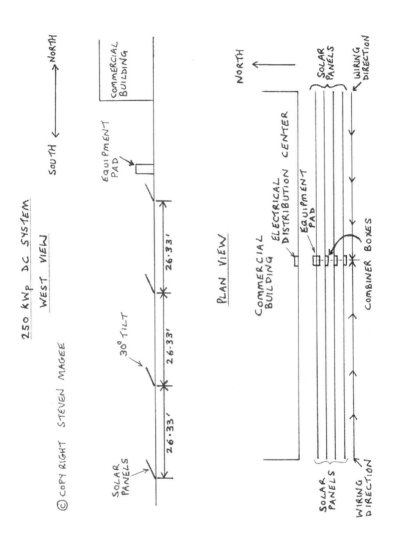

Utility Design

Utility design for the purposes of this book are systems over 1 MWp DC.

A new set of guidelines comes into play with these systems:

- Power Factor.
- AC and DC Interrupt Currents.
- Medium Voltage.
- High Voltage Transmission.
- Ride Through Capabilities.
- Harmonics Requirements.
- Large Power Swings.
- Distributed or Centralized Generation.
- Re-combiner Boxes.
- National Electric Safety Code (NESC).
- 1,000 Volt DC Systems.
- Utility Inverter Standards.
- Relays.
- Switch Yards.

Power Factor

When generating power from an inverter it is generally set to unity. However, for a utility grid, this will cause problems with the power factor regulation on the grid. Feeding in a large amount of unity power factor energy into a grid that is operating at 0.9 lagging power factor may bring the overall power factor down to a lower level, such as 0.8 lagging. It is undesirable to operate the grid at this level and as such, control of the power factor being generated by the inverter(s) is desirable. Generally utility inverters can output their power factor between 0.9 leading and 0.9 lagging. Since the grid operates near 0.9 lagging due to large inductive loads, it is suggested that you consult with the utility to obtain the power factor setting that they require. You may place stress on their power factor control equipment if you just turn on the solar inverters at the standard setting.

A nice feature of utility inverters is that the power factor can be controlled. This offers the ability of power factor correction for the grid. When designing a utility system, it is important to design in the ability to control power factor for the utility authority.

Power factor control equipment can be installed at the point of interconnection (POI) and need not be in the inverter. The advantage of this is that the inverter can generate at full power while the additional power factor correction equipment is providing the correct power factor correction to the electrical grid. Switched capacitor banks or static VAR compensators are normally used for this purpose.

VAR's reduce the amount of power that can be carried on the cables and lead to system inefficiencies. The ideal power factor is 1 but is rarely obtainable on a utility grid system due to the costs needed to produce it.

POWER FACTOR

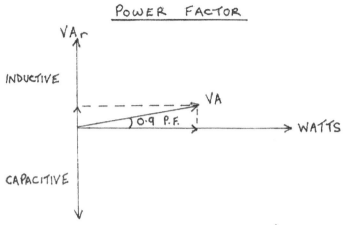

POWER COMPRISES OF BOTH REAL (W) AND REACTIVE (VAr)

WAVEFORMS

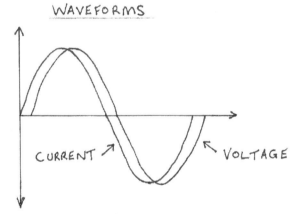

GENERALLY CURRENT WILL LAG VOLTAGE WHICH IS CHARACTERISTIC OF ELECTRICITY GENERATED WITH WOUND GENERATORS

© STEVEN MAGEE

AC and DC Interrupt Currents

At the utility level, this becomes a major concern. All fuses and circuit breakers have interrupt ratings. It is important when working with interrupt currents that you ensure that in the DC circuit that you are using the DC interrupt current ratings and in the AC circuit that you are using the AC interrupt current ratings. Generally, interrupt currents are in the range of tens of thousands of amps.

If you are using a large inverter, then it may have an equally large capacitor inside of it. It is important to find out the short circuit fault current for the inverter DC bus as you will need this to verify the interrupt current of your design. My recommendation is to use a DC design of 250 kilowatts STC or below to keep the DC fault currents to a manageable level.

You will need to ensure that you do not exceed the interrupt current of the fuses and circuit breakers used in the system. Sources of current that will need to be considered in the DC circuit are:

- Solar module maximum circuit current multiplied by the number of solar module strings connected to the inverter.
- Cable capacitance.
- Inverter capacitance.

In a inverter system of over a megawatt with several hundred strings spread over acres of land, these interrupt currents could be in the tens of thousands of amps range. Be careful when designing such large DC circuits and pay close attention to the DC interrupt current.

Long cable runs that are present on the DC circuits will generate capacitance effects. The cable manufacturers data sheet will specify the capacitance per unit length of the cabling and the total capacitance can be calculated for the cabling.

In the AC circuit, the largest fault current will generally be that from the grid. Before selecting your AC equipment, you will need to establish this value from the utility and calculate the fault currents for your particular design.

These are sources of fault currents in the AC circuits:

- The grid
- The solar power inverter systems
- Cable capacitance
- Power factor control capacitance
- Inductance

Make sure that any AC fault analysis includes all sources of power feeding into the fault. Most inverters can supply a much larger power into a fault for a few cycles. If you cannot obtain this figure from the manufacturer, then you should assume a value of three times higher than the inverter power rating.

Always calculate the fault currents for the DC and AC circuits and ensure all fuses and breakers can interrupt these currents.

The equivalent DC solar photovoltaic circuit is shown on the next page.

My book "Solar Photovoltaic DC Calculations for Residential, Commercial, and Utility Systems" has more details on calculating DC interrupt currents.

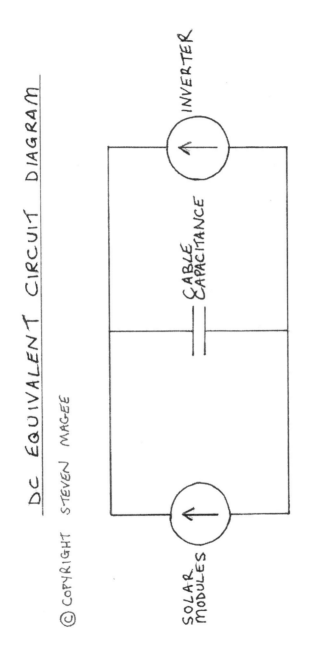

Medium Voltage

Medium and high voltage is generally classed as anything over 600 volts by the NEC. The IEEE generally classes medium voltage as over 1,000 volts and high voltage as over 60,000 volts. You will be working in these ranges on utility systems.

If you have to use poles to get your power from the solar site to the utility grid, then consider using pole mounted manual switches for maintenance purposes. These will enable you to isolate the system for maintenance purposes. This is shown in the next picture.

Distribution poles are generally grounded and on a long run, there should be at least four grounds per mile on a long run. The ground cable on the pole is shown in the following picture.

The final diagram shows how the medium voltage distribution system is put together.

Pole Mounted Switchgear

Pole mounted switches are useful for maintenance purposes.

Complete Solar Photovoltaics © Steven Magee

Distribution Power Pole

All distribution power poles have a ground wire that runs along the length of them.

Complete Solar Photovoltaics © Steven Magee

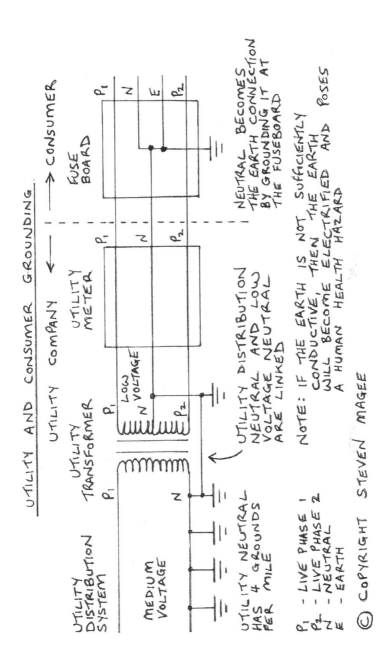

High Voltage Transmission

Transmission for the utility grid can be over one million volts. The previous designs that have been discussed operate at low voltages. Transforming the voltage up from this to medium and high voltage is straight forward. A diagram showing the utility transmission system is on the next page.

You will need to know the following items to design the transmission interconnection:

- Interconnection voltage.
- Interconnection power capacity.
- Wye (star) or delta feed into the distribution transmission network.
- Does the interconnection have a current carrying neutral and/or ground conductor?
- Interconnection short circuit fault current.
- Transformer impedance.
- Harmonics requirements.
- Protection relay system required by the utility.
- Protection relay settings required by the utility.

With these, you will be able to design the solar photovoltaic system interconnection. It is important when designing the interconnection that you use electrical switchgear that is capable of being back fed in your design.

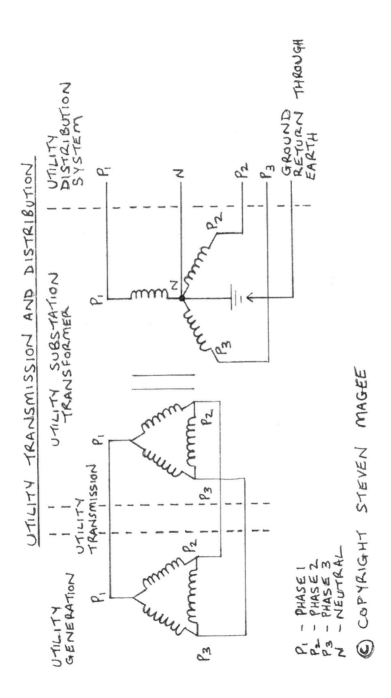

Ride Through Capability

Utilities want to generate power and if the grid voltage dips, they want their power generation systems to stay on line and "ride through" the dip. This is a different concept than previously discussed and is unique to the utilities. Low voltage ride through (LVRT) typically lasts for nine cycles down to 0 volts.

It is important when building a system for a utility that you select a utility grade inverter that has this feature. The utility wants to keep power feeding into the electrical grid and wants the grid to stay on line. Generally the ride through feature can be set to keep feeding into the grid for a specified time before the inverter will shutdown. Check with the utility what their requirements are for this feature.

Complete Solar Photovoltaics © Steven Magee

Harmonics Requirements

Most utility inverters generate power at below 3% harmonic distortion. This is generally acceptable by the utilities and multiple inverter systems can keep the harmonic distortion below this level. When selecting your inverter system, it is always good to get confirmation from the manufacturer that a large, multiple inverter system can keep within this tolerance.

If you look at the waveform on the next page you will see that it is a 60 Hz sine wave.

If you look at the same waveform that is on the utility system with an oscilloscope with a Fast Fourier Transform (FFT) you will see that it looks like the image on the following page. It is normal to see the odd harmonics as they are a feature of all electronic power generation and electronic loads in general. In the image from left to right you can see the 60 Hz (1st), 180 Hz (3rd), 300Hz (5th), 420 Hz (7th), and 540 Hz (9th) harmonics as spikes. The even harmonics generally cancel out on a utility system.

If you have really severe harmonics on the system you will need to start troubleshooting it. Your utility sine wave should never look like the last picture! A waveform like this may make the wiring on the utility system to radiate radio waves!

Complete Solar Photovoltaics © Steven Magee

Utility 60 Hz Sine Wave

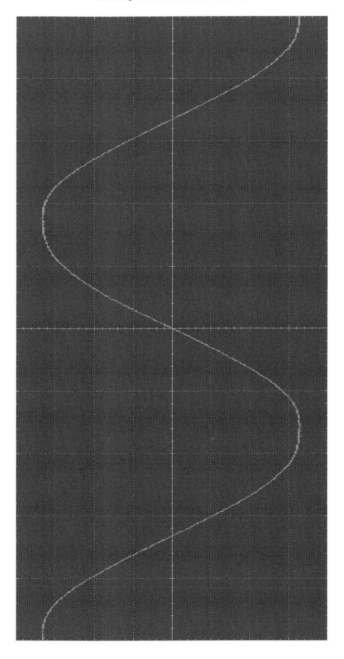

Inverter Harmonics

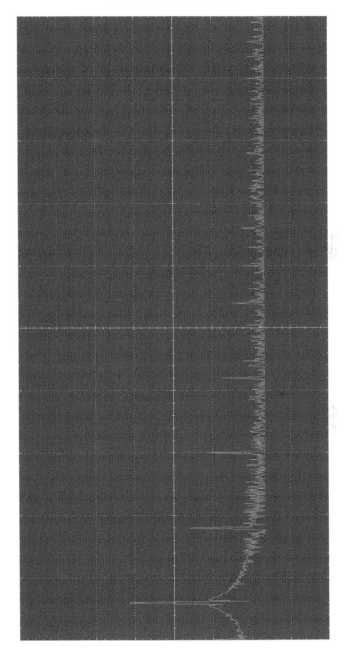

Complete Solar Photovoltaics © Steven Magee

Severe Harmonic Distortion Sine Wave

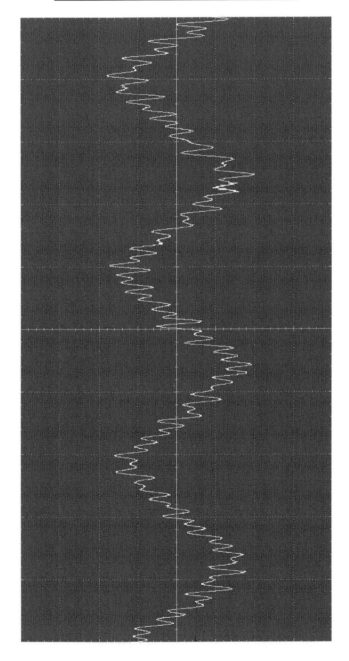

Large Power Swings

Any solar photovoltaic system will suffer from large power swings that are proportional to irradiance. Solar cells directly convert irradiance into energy in real time. The main thing that can significantly effect the power output quickly are clouds. Clouds can increase the power as well as decrease the power output. Broken clouds will reflect and lens light from the Sun and increase the power output from the solar modules that are in direct sunlight. As the cloud starts to approach the Sun, lensing will occur that will focus more sunlight into the solar module. This surge in power can be a ramping up of normal power levels that produces up to a 50% increase for a few minutes. As the cloud progressively covers the Sun, the power will start to drop off significantly with drops of over 90% from normal power levels. On a broken cloud day, these large power swings can occur within several seconds if the cloud is fast moving or over several minutes if they are moving slowly.

As you can imagine, these large power swings cause havoc with the grid and can look like faults to the grid control system. On a large power generation system, it is important to build into the grid control system feedback from the solar power control system so that the grid can discern between a genuine fault and clouds passing over the solar power generation system.

The grid needs to be able to tolerate these large swings and alternate conventional power generation needs to be able to rapidly react to the power surges that the solar power system will generate during broken cloud days.

These effects can be seen on the following diagram. The top graph shows the system power output on a clear day. The bottom graph shows the power output on a day with broken clouds. Peak irradiance is achieved on the broken cloud day due to the cloud effect.

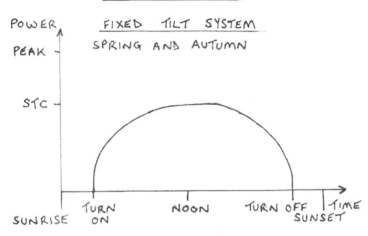
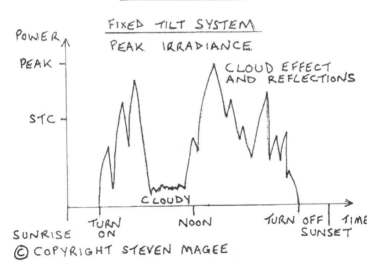

Distributed or Centralized Generation

Distributed power generation refers to when the solar power generation system is broken into many small parts and spread over a wide area. Residential installations and commercial installations would fall into this category. Distributed power generation is less prone to broken cloud effects due to the solar modules being spread over a large area. On a broken cloud day, some solar power systems will be in the Sun while others will be in the shade. It creates an averaging of the broken cloud effect on the grid and is much more desirable.

Centralized power generation is when a large solar power generation system is installed in one location. Utility scale installations would be in this category. There are advantages regarding maintenance costs to centralize the utility solar power plant, but these seem to be negated by the cloud effects on the solar power system output. Centralized power generation works well when complimented by another source of fast response power generation, such as a gas turbine power generation plant. Thus the solar power generation system is used to offset the fuel consumption of the conventional power generation plant.

Re-combiner Boxes

Re-combiner boxes are used to combine the outputs of two or more combiner boxes into a larger cable. They generally will have two to four large circuit breakers inside them, depending on the amount of combiner box circuits being combined.

A re-combiner box wiring diagram is shown on the next page.

Complete Solar Photovoltaics © Steven Magee

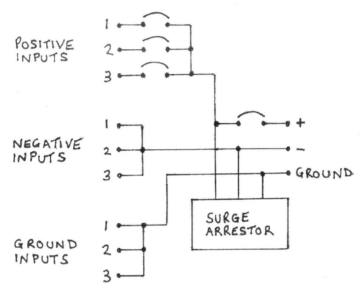

National Electric Safety Code (NESC)

Now that we are on the utility side of the meter, a new standard applies. The National Electric Safety Code (NESC) covers the utilities. One thing that you will notice is that there is no mention of solar photovoltaics in the National Electric Safety Code. Due to this, we obtain our solar photovoltaic design equations from the National Electric Code (NEC). The National Electric Safety Code (NESC) and the National Electric Code (NEC) should be read in conjunction with each other. You will find on utility scale installations that you will refer to the two code books frequently.

1,000 Volt DC Systems

There has been a move towards 1,000 volt DC systems in the USA. 1,000 volt systems are widespread in Europe. The advantage is that the efficiency of the DC circuit is increased due to lower power losses in the circuit. Cabling and equipment costs are reduced as less is used. I personally do not recommend these systems as a solar photovoltaic systems need maintenance at least annually and this high voltage increases the electrocution risks for the workers performing this maintenance. National Electric Code section 490.2 defines high voltage as over 600 volts. As such the provisions in National Electric Code section 490 as well as the National Electric Safety Code (NESC) should be followed.

Utility Inverter Standards

FERC 661-A is a wind-power standard that is typically being applied to the utility inverters. IEEE 1547.8 is also being developed for the utility inverter industry. You should get familiar with these standards if you are working in the utility solar power industry. The utilities want their inverters to have volt amperes reactive (VAR) control, low voltage ride through (LVRT), and dynamic command and control.

Relays

Relays monitor the grid and control the switchgear. There are many trip types and sequences for the switchgear to operate in. The relays will trip the system off if over-voltage or under-voltage is detected on it. Over or under frequency will do the same thing. High harmonics may also trip the system. Relaying is an expert field and it takes many years of experience to set up relays so that the system provides reliable generation and acts correctly during the various fault conditions that may be presented. Unfortunately, when incorrectly set up, they can take down the large portions of the electrical grid.

Switch Yards

On a large system you may well find yourself installing a high voltage switch yard. You should be familiar with the design of these and their requirements. Be aware that anytime that you enter a switch yard that there are voltages and curents present that can be fatal. There can be several hundred thousand volts present within these.

A typical switch yard is shown on the next page.

Typical Switch Yard

You should be very careful designing and working around equipment such as this. Many people are killed inside of these.

10 MWp DC System

We will now design our utility scale system. For this design we will use 250 kW inverters. Our DC circuit, the mounting system and inverter design is the same as the commercial design from the previous chapter. We now have to connect 40 inverters together and feed the power into the grid. This is a relatively straight forward task and can be done in many ways, all of them would be equally suitable.

This particular design will consist of four 2.5 MW transformers feeding into distribution centers that connect to ten inverters each. This gives us a segmented design that during a large scale equipment failure, such as a faulty transformer, we should only see 25% of the power drop off from the grid.

Our commercial solar modules and inverter system are suitable for this design and we will now use that again here for each of our 250 kW systems.

We will end up with a grid of solar power systems on this design covering about 100 acres of land. We will build four blocks of solar, each a 2.5 MVA system.

We will take the outputs from each of the four systems, combine them using a medium voltage distribution center and the utility will connect them into the 24,000 volt utility poles that pass along the edge our piece of land.

Dry type, high efficiency power transformers are generally used in solar power applications. These have efficiencies of over 99% and are low maintenance. We will use these for our application.

Both our AC low (480V) and medium voltage (24,000V) circuits will be protected by ground fault detection systems. These will use protective relays to detect ground faults. The ground fault current will be limited in these systems to a low level. This will protect the cabling from serious damage from ground faults.

Now that we have our system specifications, we will now run our calculations to design it. Our design is the same as the commercial design from the previous chapter and we will now expand that design to utility scale.

We will combine these ten inverters using an electrical switching center. This will give:

Switching center output current = 10 x inverter current

= 10 x 301 amps

= 3,010 amps

Our minimum fuse size is 125% larger than the circuit current:

Cable amperage = 125% x 3010A

= 3,762.5A

NEC240.6(A) indicates a standard fuse size of 4,000A for this circuit.

This seems like a lot of current, but we are traveling just a short distance to the transformer that is mounted on the same equipment pad, so our cable length is minimal. Let's see how this compares with our de-rated values:

De-rate the current for an ambient temperature of 40°C:

De-rated current = 4,000A x 0.91

= 4,396A

Looking at NEC table 310.16 we see that 90°C 2000 kcmil copper cable can carry 750 amps. So we will need to parallel up the cables to achieve the desired current:

Number of parallel cables needed = de-rated current / 750 amps

= 4,396A / 750A

= 5.87

We round up to 6 parallel cables of 90°C 2000 kcmil.

Cable current = Switching center output current /6

= 3010A / 6

= 502A per cable

We feed into the transformer 480 volt winding. On the output we get 24,000 volts. This gives:

Transformer step up = 24,000V / 480V

= 50

Since we have increased the voltage by 50, we have correspondingly reduced the current by 50. We also need to allow for the efficiency of the transformer at 99%.

MV current = Switching center current x 99% / 50

= 59.598 Amps

This gives an MV power of:

MV transformer power = sqrt(3) x 24,000V x 59.598A

= 2,477,443W

We come out of the transformer to a MV fused switch. We apply 125% to the current to get:

MV fuse size = 125% x MV fuse current

= 125% x 59.598A

= 75A

NEC240.6(A) indicates a standard fuse size of 80A.

We would use the larger de-rating of either 125% or the de-rating for a 60°C temperature if the fuse manufacturer requires it.

80A is our fuse size and also our minimum cable size. Our cable will be in an underground duct, so we will look into using the following cable per NEC table 310.77:

MV Cable = 90°C MV90 6AWG copper cable = 90A

This is rated at 20°C ground ambient temperature. We are using 40°C for the ambient ground temperature, so let's check on that for de-rating using formula NEC 310.60-C-4:

Ambient current= cable rated current x sqrt((Conductor temperature - desired cable ambient temperature - dielectric loss temperature rise) / (Conductor temperature - cable ambient temperature from table - dielectric loss temperature rise)

= 90 x sqrt ((90 - 40 – 20) / (90 - 20 – 20))

= 70A

As we can see, this cable is not large enough for our purpose after de-rating, so we move onto the next cable size. This is 4 AWG 90°C type MV 90 copper at 115A.

= 115 x sqrt ((90 - 40 – 20) / (90 - 20 – 20))

= 89A

So 4 AWG is our choice for the 24kV 2.5MW cabling.

Now onto the 10MW 24kV cabling.

10 MW current = 10MW / sqrt(3) x 24,000V

= 240.57 Amps

To obtain our cable size, we apply 125% to the current to get:

10MW fuse size = 125% x MV current

= 125% x 240.57A

= 301A

NEC table 310.77 indicates that 250 kcmil 90°C MV90 copper cable can carry 325A. Now we de-rate this for an underground temperature of 40°C to obtain our cable size.

= 325A x sqrt ((90 - 40 – 20) / (90 - 20 – 20))

= 251A

As we can see, this cable is not large enough for our purpose after de-rating, so we move onto the next cable size. This is 350 kcmil 90°C type MV 90 copper at 390A.

= 390A x sqrt ((90 - 40 – 20) / (90 - 20 – 20))

= 302A

So 350 kcmil is our choice for the 24kV 10MW cabling.

Our switchgear in the AC circuit will be selected based on the interrupt current supplied by the utility. The equipment grounding will be done in accordance with the equipment manufacturers recommendations and local codes.

So here are our system specifications:

- 10 MVA Cabling: 90°C 350 kcmil type MV-90 copper at 24 kV
- 1 of: 10 MVA, 24 kVAC Distribution center with ground fault detection
- 2.5MVA Cabling: 90°C 4AWG type MV-90 copper at 24 kV
- 4 of: 2.5MVA, 24 kVAC medium voltage pad mounted fused switches with 80A fuses
- 4 of: Medium voltage transformers: Dry type, 99% efficiency, 2.5 MVA, 24 kVAC / 480VAC
- 4 of: 2.5 MVA 480VAC Switching centers with ground fault detection and ten 400A circuit breakers
- 480 VAC Switching center cables 90°C 2000 kcmil copper
- 40 of: Inverter 250 kW, 3 phase, 480 VAC, 600 VDC with internal transformer
- Inverter 480 VAC cabling: 90°C 500 kcmil copper
- Inverter DC cabling for 10 input combiner: 90°C 1/0 AWG copper
- Inverter equipment ground cabling for 10 input combiner: 90°C 6 AWG copper
- 160 of: 10 input combiner boxes

- Inverter DC cabling for 9 input combiner: 90°C 1 AWG copper
- Inverter equipment ground cabling for 9 input combiner: 90°C 6 AWG copper
- 160 of: 9 input combiner boxes
- Solar string cabling: 90°C 14-10 AWG copper
- 3,040 of: 15A DC solar photovoltaic string fuses
- 42,560 of: solar modules
- Row Spacing: 26.7 feet

When the utility provided the interconnection details you will be able to calculate the fault currents in the design. With these values the correct switchgear can be selected, ensuring that it is rated for back feed.

Once the design is completed, it will need to be stamped by a licensed Professional Engineer to verify the design and the calculations.

The following pages show our design:

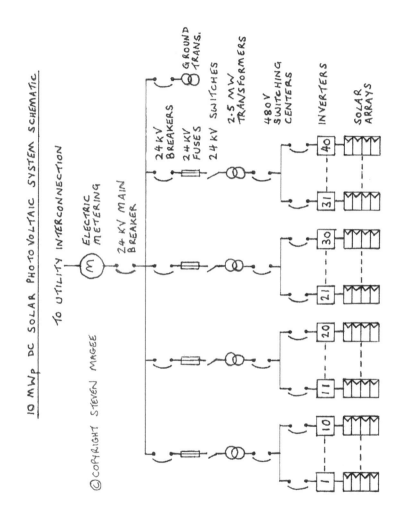

10 MVA DISTRIBUTION

© COPYRIGHT STEVEN MAGEE

- UTILITY INTERCONNECT — 3-PT, 3-CT — 1
- METERING — 1-op. tr., 3-PT — 2
- 2.5 MVA TRANSFORMER — 3-CT — 3
- 2.5 MVA TRANSFORMER — 3-CT — 4
- 2.5 MVA TRANSFORMER — 3-CT — 5
- 2.5 MVA TRANSFORMER — 3-CT — 6
- GROUNDING TRANSFORMER — 3-CT — 7

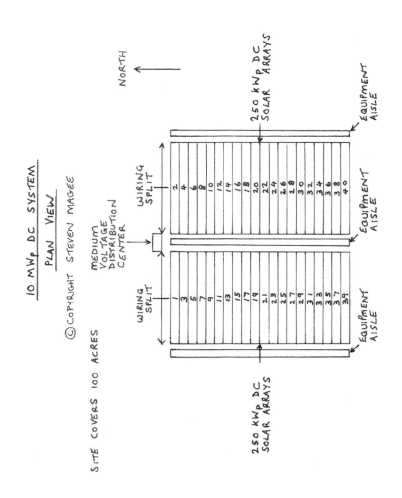

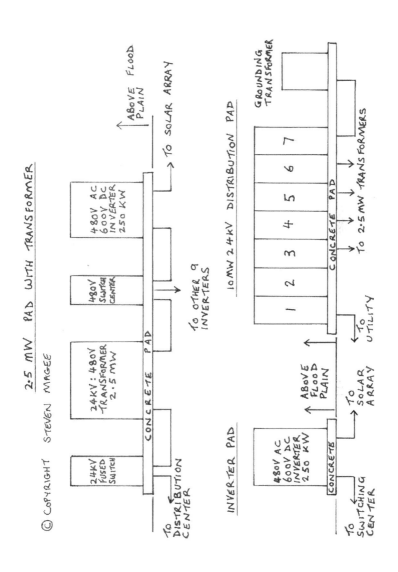

Construction and Commissioning

Once you design has been submitted and approved, you are ready to mobilize.

Notify all of your suppliers that you are starting the project and secure guaranteed delivery dates for all equipment. Once you have these lead times, you can start working with your contractors to firm up a schedule for the construction of the project.

A wise project manager will build into the schedule contingency for weather delays, unforeseen problems, equipment delivery problems and so on. The important thing with a project is to ensure that it is built correctly and built well. Many a hare has built projects that have turned out to be lemons because they rushed construction.

If you can, build out the project in sections that can be commissioned as the next stage is being built. A logical process for construction is:

- Survey project site for grading.
- Grade project site.
- Survey project site for construction.
- Fence project site.
- Build medium and high voltage distribution and transmission system.

- Install weather monitoring and security systems.
- Build inverter 1 system
- Commission inverter 1 system and construct inverter 2 system.
- Incorporate lessons learned from inverter 1 system into the rest of the site plans.
- Commission inverter 2 system and construct inverter 3 system.
- Keep this process flowing until the system is completely built out.
- Incorporate all changes into the drawing set and mark them "As Built".

Slow and steady leads to well built solar power plants. I am always amazed at the rapid speed that companies want to construct these large utility systems at. There generally is no need to rush these big projects and as the designer you should be querying the project schedule if it appears unreasonable.

The ideal is to build one full inverter system and run it for one year before building the rest of the site out. There are many problems that may appear during the first year of operation that may save money by being built into the rest of the inverter systems at the start rather than the end. All sites have a certain level of uncertainty associated with them until a solar power system has been in operation on them for at least one year.

For roof mounted systems and other systems with poor access, you should verify that each module is functional before installation due to the difficulties in troubleshooting these systems. Simply expose each of them to the Sun and ensure that they are producing the correct voltage with a multimeter prior to mounting it. If you have the competence, you could also do a short circuit test by connecting the positive and negative leads together while the module is covered with an opaque cover. Use a DC clamp meter to check the current from it when exposed to the Sun. Cover up the module with an opaque cover to undo the leads prior to installation.

You should be aware that some of the photovoltaic technologies actually produce more power than their rating for the first few months of service. They have a "burn in" period before they settle into their normal power rating. If you initially turn on a system and it is producing too much power, check with the manufacturer about this.

The following pictures show some of the aspects of constructing the larger systems.

Complete Solar Photovoltaics © Steven Magee

Construction of System

Construction of a large single axis tracking system is shown.

Complete Solar Photovoltaics © Steven Magee

Staging Area

You will need a staging area during construction to hold the large number of items needed to build the project. These are the solar modules that are awaiting installation.

Site Office

You will be working in a site office during the construction of the project.

Tools Needed

The tools needed depend on the type of solar photovoltaic work that you are doing. They get more specialized as you increase the size of the systems that you are working on. In general, your starting point is a standard electrician's tool kit with a few extra items.

This is a basic residential and small commercial set of tools:

- Infrared digital camera with memory card.
- Oscilloscope with Fast Fourier Transform (FFT).
- Trifield meter.
- Irradiance meter.
- Thermometer.
- Wind speed meter.
- AM Radio.
- DC and AC clamp on ammeter.
- Multimeter with min/max logging.
- Insulation test meter.
- Solar connector termination tools.
- Cordless drill
- Cordless screwdriver
- Cordless impact driver
- Drill bits for wood and masonary.

- Insulated screwdrivers.
- Insulated pliers.
- Allen wrenches.
- Socket set.
- Hammer.
- Sledgehammer. (for ground rods)
- Rubber mallet.
- Insulated gloves.
- Insulated steel toe boots.
- Hard hat.
- Eye protection.
- Ear protection.
- Fall protection.
- Ladders.

The infrared camera is the most expensive piece of equipment on the list and is essential for proving warranty claims on modules with the manufacturers. You will need the oscilloscope for diagnosing faulty inverter systems. The AM radio and Trifield meter are useful for diagnosing electromagnetic interference (EMI) problems that customers may be complaining about.

As a precaution you should use your test equipment to establish the levels of EMI that the equipment that you are working on produces and limit your time in high EMI environments. EMI does strange things to people! Wise solar engineers use known low EMI producing equipment to build systems with. You should notify the manufacturer and the government if you find equipment that is producing high amounts of EMI.

As you progress onto larger commercial and utility installations you will need:

- Infrared video camera.
- Time domain reflectometer (TDR).
- Three phase harmonic analyzer.
- Logging multimeter.
- Hipot test meter.

The utility test equipment can be quite expensive!

General Problems

There are a number of problems that are common to all pieces of the solar power system. These are:

- High thermal cycling.
- Physical damage.
- Wind and snow loading.
- Frost damage.
- Corrosion.
- Fire and overheating.
- Theft.
- Weeds and vegetation growth.

High Thermal Cycling

On sites that have large extremes between the daytime temperature and nighttime temperature, you should expect to see many faults that are related to thermal cycling problems. The constant expansion and contraction of components may prematurely wear them out. Bad connections and loose hardware are generally the problems that you will see. If you are constantly tightening screws and nuts on your system, you should be suspecting thermal cycling problems.

Physical Damage

Physical damage can occur from storms. Any time a storm has passed through, you should give your system a check. Branches and debris that get blown into the system may cause damage to solar modules, their wiring, and the system in general. Frame-less solar modules are more prone to this type of damage.

The racking systems can get physical damage from high wind speeds and from inattentive drivers and you may need to hire specialized equipment to repair it. Racking systems may also corrode through at their bases which may need the same type of repair.

A typical racking post diver is shown on the next page.

Racking Post Driver

The racking posts are typically driven in with a hydraulic post driver.

Wind and Snow Loading

If mistakes were made in the structural engineering of the solar power system, you should not be surprised to find that your system has literally blown away in high winds! Ground mounted systems may be bent over by the force of the wind. Generally, you would only see this type event in a severe storm in the area.

Snow loading can destroy a system if not designed correctly. The glass on solar modules may crack and racking systems may collapse. A severe snow storm can add a lot of weight to the system.

Frost Damage

Frost damage may cause solar module glass to crack, foundation problems, and may damage equipment in the field. The expansion of freezing water that has made its way into gaps and cracks is what causes the issues.

Corrosion

Corrosion may be a serious issue if the system designer did not understand galvanic corrosion effects. This occurs when dissimilar metals are bolted together. It takes time to occur and when it does it may be throughout the entire system. Identify the parts that are incompatible and replace them with the correct parts.

Galvanized steel cuts will show corrosion effects if they were not treated during installation. You should ensure all cuts in galvanized steel are coated properly.

Racking Systems

Racking systems are built quickly and you may find that you are having corrosion problems later due to the inattention of the installers.

Fire and Overheating

Fires and overheating can be a problem on solar photovoltaic systems. De-rating of components and the use of outdoor equipment can cause problems if they were not accounted for. If you have a fire on your system, then you should check through the design to see if mistakes were made during the design stage. If you notice components overheating, then you should do the same.

You should not open or break any DC connection when the system is in operation. You may get shocked or electrocuted and you may start a fire. You will most certainly cause damage to the equipment by doing this. DC will arc and the arc may not extinguish until sunset when the system stops producing power. The only DC connections that should be opened under load are the DC disconnects.

Parallel DC arcs will occur when the positive and negative DC connections are shorted. Series arcs will occur when DC connectors are pulled apart under load or when DC fuses are removed under load.

Theft

Grounding systems are the most common piece of electrical equipment to be stolen. On any maintenance visit it should be the first thing that you verify. If it has been stolen, then your system may be in a dangerous state and you should exercise caution around it. It should be taken out of service until the grounding system can be repaired.

Solar module theft has not been common and does not appear to be an issue currently. It probably reflects the lack of demand for industrial solar modules, there are no buyers for this type of equipment. Solar modules are firmly attached and need special tools to remove them which most thieves do not have. If you are having modules stolen then you should install anti-theft mountings on the modules.

Most theft is centered around the copper and aluminum cables which have a high value as scrap materials.

Keep a regular check on the the fencing system for the site and repair any damage to it promptly. Make sure that a sign is displayed with contact information on the entry to the fenced area and the usual legal signs for trespassing and electrocution risks.

Weeds and Vegetation Growth

You will need to stay on top of your vegetation growth around equipment. Weeds will block ventilation on equipment, shade solar modules, and present a fire risk during droughts. While weeds are commonly a problem for ground mounted equipment, they can also be an issue for roof mounted equipment. It is not unusual to see weeds growing in roof mounted systems!

Solar Module O & M

Solar modules typically represent the largest investment in the system. Traditionally silicon solar modules have been used, but now many newer types of technologies are emerging such as thin film, and so on. All solar modules are tested to the Underwriters Laboratory (UL) 1703 standard in the USA.

Crystalline silicon is the best understood, the most efficient, and has been around for many decades. The newer film type solar modules are less efficient and cheaper to purchase. Unfortunately more thin film modules are needed to generate the same power as silicon, and this increases the system physical size and associated support systems such as cabling, racking and so on. The larger the physical system size the more the operation and maintenance costs will be.

Generally any decision on which technology to use is driven by market rates for each type of technology, aesthetics and personal preference. Solar modules are a commodity and their prices can fluctuate rapidly. As such, your system may be built from any of the different solar module types.

There are two types of photovoltaic module construction in use. The difference is the module backing. The modules that have a glass backing are typically stronger. Modules that have a flexible back sheet are cheaper to make. Most of the cheaper solar modules will have the flexible back sheet. Frame-less modules and high wind load modules tend to have the glass backing.

To understand the solar module failures, we need to understand how a solar photovoltaic module is constructed. The individual solar cells that you can see through the glass are actually very large diodes. When these are exposed to light, they generate a low voltage, typically 0.6 volts per solar cell. This is too low to use, so it is increased by connecting the cells in series with each other, this is called stringing. This increases the voltage to a higher value, typically about 10 volts DC per internal string.

There are generally two to four internal strings of solar cells inside the modules. They are connected together in series through the junction box on the module to raise the voltage even higher. The voltage value shown on the solar module electrical specifications is obtained by doing this. Typically this will be between 20 volts and 60 volts DC, depending on your particular solar modules.

Inside the solar module junction box there are diodes. These are called bypass diodes. The function of these is to bypass any internal solar module strings that may not be producing sufficient current due to either an electrical failure or shading of the cells. If everything is okay electrically with the solar module and it is not dirty or shaded, then these are not generally in use.

A number of solar modules are connected in series to increase the voltage further. This is called a string. This is the final increase in DC voltage and can be up to 600 volts on a residential or commercial system and up to 1,000 volts on a utility system.

Solar modules are very reliable and rarely fail. The failure rate is often quoted as less than one percent annually. The typical solar module failures are:

- Manufacturing defects.
- Broken glass.
- Faulty bypass diode.
- Failed solar cell.
- De-lamination of the solar module.
- Backing sheet damage.
- Bad connections in the junction box.
- Bad connection within the solar module.
- String wiring problem.
- Dirt.
- Bad grounding.
- Lightning damage.
- Voltage overloading.
- Insulation faults.
- Reverse current.
- Aging effects.

Let's look at each one of these in detail:

Manufacturing Defects

Modules have failures that are related to their production run. Certain production batches have high failure rates. If you are seeing a lot of failures, then it may be due to this. Check with the manufacturer to rule this out.

It appears that certain types of thin film modules may have higher failure rates than crystalline silicon modules and this may be a reflection that the technology is relatively new and has not yet stabilized. Thin film products are more prone to damage from the effects of moisture.

Broken solar cells in the module can be seen through the glass and these are a manufacturing defect. If you have any in your system and the module fails prematurely, it should probably be replaced under warranty.

Broken Glass

Broken glass can occur for many reasons:

- Physical damage to the glass
- Large hail
- Thermal shock
- Improper manufacture
- Improper installation

Complete Solar Photovoltaics © Steven Magee

There isn't much that can be done about physical damage. If someone decides to throw objects at the solar modules, then they may break. High winds can also kick up debris and sent it towards the glass. The glass is very tough on the modules and usually they will not break. Hail would have to be very large for it to have any effect. Wind born sand and small stones may scratch the glass surface and weaken it. Grounds maintenance activities may cause stones to be thrown onto the modules from grass cutting and weeding.

An important concept to realize with solar modules is that once the glass is broken, the electrical insulation of the solar module may be compromised. This could mean that there may be the full string voltage available at the surface of the glass, particularly if it is wet. This could be up to 1,000 volts DC on a utility system. Take care when working on modules that have physical damage. Any damaged module should be physically removed from the system as soon as possible due to this.

An electrical fault internal to the solar module can create heat. When this happens in the case of a faulty solar cell, it can actually start to change color due to the heating effects. If you are seeing a different color solar cell to the rest of the cells inside the solar module, then you likely have a hot solar cell. The effect of this heat is that the glass in front of the hot spot will start to be heated. If this heat is too excessive, then the glass will break. It is no surprise that glass breaks on solar modules more frequently in the winter time than summer time, just like car windscreens. The thermal shock effect is greater in wintertime due to frosts.

Manufacturing faults may cause failures. If you are seeing a lot of glass failures then it will most likely be either a manufacturing fault on that batch of solar modules or poor installation. It is worth checking with the warranty department of the manufacturer to rule this out.

Poor installation will break solar modules. Generally the solar module may not be mounted in accordance with the manufacturers installation guide and this will put stress on the frame and the glass. If this stress is too much, then eventually the glass will break. If you are seeing a lot of solar module glass breakage, you should check the mounting system and the solar module clamps. It is common for installers to over-torque the module mounting hardware and to put twists into solar modules from uneven racking installation.

Glass breakage from electrical faults may be covered by warranty. The glass tends to have a center breakage point that can look somewhat like the wings of a butterfly and the breakage radiates out from that point. If you have a glass breakage problem that you can not trace to physical damage, then you may have a warranty claim. Consult with the manufacturer about glass breakage warranty claims.

The typical glass breakage pattern that the warranty should cover is shown on the next page. You should spend time looking for the "Butterfly" effect when replacing broken modules.

SOLAR MODULE GLASS BREAKAGE

BREAKAGE THAT OCCURS WITH THE "BUTTERFLY" EFFECT IS TYPICAL OF A "HOT SPOT" AND MAY BE COVERED BY WARRANTY.

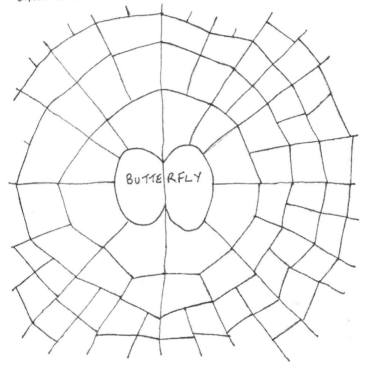

WHENEVER YOU FIND A BROKEN MODULE YOU SHOULD LOOK FOR THIS PATTERN IN THE BROKEN GLASS. (NOT TO SCALE)

© COPYRIGHT STEVEN MAGEE

Faulty Bypass Diodes

A shorted bypass diode will do two things. It may get hot and it will short out the internal module string of solar cells that it protects. This will show itself in two ways. The first is to possibly melt the junction box on the solar module. The second is to short out the internal string of solar cells. This would show itself as a drop in voltage on the module corresponding to how many solar cells were shorted out by the faulty bypass diode.

An open circuit bypass diode is harder to detect. The bypass diodes are used for shading effects and to bypass electrically faulty internal strings or solar cells. If the internal string of solar cells has gone open circuit and the bypass diode is also open circuit then no power will be generated by the solar module nor the string of modules that it is connected to. Change out the faulty solar module if this is the case.

For shading effects a current test would have to be performed, shading each internal string on the module separately and seeing if the current drops significantly. If it does, then you will have detected a faulty bypass diode. Replace the diode with the same type.

The solar module junction box contains the bypass diodes and is shown on the next page.

Complete Solar Photovoltaics © Steven Magee

Solar Module Junction Box

The solar module junction box contains the bypass diodes.

Failed Solar Cell

The solar cell can fail in a number of ways. It can go open circuit, short circuit, or hot. If it opens, then the bypass diode will bypass all current from the internal string of cells that it is connected to. This failure can be seen as a lower voltage on the terminals of the solar module, corresponding to the number of solar cells that are in the faulty internal string.

If the cell shorts, then the individual cell will probably will be cool. A shorted cell would be seen as a small voltage drop of about 0.6 volts on the solar module terminals.

A hot cell is exactly this. The cell produces less current than the others in the string and starts dissipating the excess current passing through it from the other cells as heat. It may look discolored visually.

A thermal infrared camera is very useful in diagnosing these faults.

Unfortunately, for failed internal cells there is no repair. Replacement of the solar module is the only solution.

The hot cell is shown on the next page.

HOT SOLAR CELL

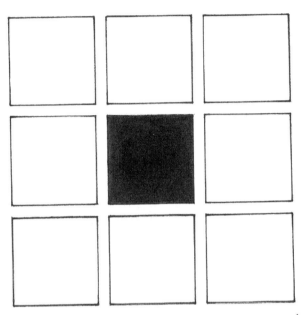

WHEN IMAGING A SOLAR MODULE WITH AN INFRARED CAMERA, THE "HOT CELL" WILL SHOW UP AS A DIFFERENT SHADE TO THE REST OF THE CELLS. SOMETIMES YOU WILL BE ABLE TO SEE A "HOT CELL".

© COPYRIGHT STEVEN MAGEE

De-lamination

A common manufacturing problem is de-lamination of the solar module. De-lamination is a failure of the sealing of the solar module. The front glass of the solar module separates from the backing sheets or backing glass. When this occurs, water can start to ingress the modules and failure is likely to occur afterwords. This can be seen as corrosion of the internal solar module wiring and internal solar module discoloration over time. Module replacement is the only solution for this.

Backing Sheet Damage

Damage to the solar module backing sheet can be a common problem. Once the backing sheet is damaged, then the module electrical insulation has been compromised and should be replaced.

Bad Junction Box Connections

Bad junction box connections will either cause open circuits or electrical arcing. In the case of an open circuit, the module will not function. Electrical arcing is a concern as it will start to damage the junction box and may send heat into the solar module. Generally, electrical arcing will show itself with a melting junction box.

Both of these failures are repairable by the module manufacturer and the module should be returned for repair.

Internal Solar Module Bad Connection

The solar cells are connected in an internal string by wires. If the soldering is faulty or the wires break, then the internal string will cease to function. The bypass diode will then bypass this faulty internal string of solar cells. This will be seen as a lower voltage corresponding to the number of solar cells in the faulty internal string.

If current is flowing in the module and the bad connection does not break the circuit cleanly, then it may start to arc and create heat. In this situation you would expect the glass to break and a possible fire to occur from the arcing.

The bad connection may just run hot and this would create a hot spot on the surface of the module. The backing sheet may develop a brown mark or a burn where the overheating is taking place.

This fault cannot be fixed and the solar module will need to be replaced.

String Wiring Problem

The solar modules are connected into strings by the string wiring to increase their voltage to that of the DC system voltage, generally a few hundred to one thousand volts DC. The string wiring and connectors can fail in either an open circuit connection or an arcing connection. If the string is showing open circuit, then check the wiring and connectors. If arcing, disconnect the DC string using the

DC disconnect switch, remove the string fuse and repair the problem.

A common problem is badly terminated connectors. If you are seeing a lot of connector failures, it may be that they were terminated incorrectly. It would be prudent to inspect all connectors if this is the case.

Dirt

Dirt can severely affect the power output of a solar module. If they are visibly dirty, then it is probably time to clean them to return them back to good performance. Quite often solar modules look clean until you get close to them and can see the dirt and dust on them. When checking a system for dirty solar modules, you should clean one module and see how it compares to the rest.

A great way to get electrocuted is to wash solar modules by hand during the day time on grid interconnected systems. If the module is cracked or has an electrical insulation fault and you throw water onto it, the water could rise to the voltage of the solar module string, possibly 600 to 1,000 volts DC. Follow the manufacturers instructions for cleaning the modules and, if possible, use a non contact washer that atomizes the water. It is inadvisable to come into contact with a wet solar module that is producing electricity. Even if the DC disconnect is opened, the solar modules still have voltage on them and the capacity to supply current. Night time cleaning is recommended for this reason. Wearing protective equipment such as electrically insulated gloves and boots will help reduce the risks.

Solar modules can produce electromagnetic interference (EMI) effects when in service and prolonged contact with solar modules that are in service may make you ill.

You should not pressure wash solar modules as you may invalidate the warranty. Consult with the manufacturer of your solar modules to verify the correct way to clean them.

Modules that spend their time in an orientation close to horizontal may get excessive build up on them relatively quickly. This is shown in the following picture.

Complete Solar Photovoltaics © Steven Magee

Horizontal Solar Modules

Horizontal solar modules build up the dirt the quickest. You may be frequently washing a system with modules mounted close to horizontal.

Bad Grounding

All solar module frames should be electrically grounded. If a module with an insulation fault on it is improperly grounded, then this can cause arcing between the frame and mounting system. If there are any signs of this, replace the solar module and check the equipment grounding system.

Always inspect the solar module equipment grounding system and tighten the connections during maintenance.

Lightning Damage

Lightning damage can usually be seen as a burn mark on the solar module, generally on the metal frame. Normally the module will be damaged beyond repair and it will need to be replaced. Sometimes if the lightning strike had a large amount of energy contained in it, this damage will be in the string of modules. Damage should be contained to the string. Again, replace all failed modules to repair. The fuse for the string will likely need replacement. The string fuses and low resistance grounding are generally a good protector against extensive system lightning damage.

If you are getting a lot of lightning damage on an installation, you may want to consider cleaning up the cabling on the installation to remove loops that may have been formed in the cabling.

Voltage Overloading

If the designer of the system did not factor in temperature effects for the solar module and adjust the voltage accordingly, it is quite possible that you may find that your system is overloading in voltage with the change in seasons. While hopefully this should not cause any damage to the solar modules or inverter, it is undesirable. If you find that your system has this condition, it is easily fixed by reducing the number of solar modules in the string(s) to an acceptable DC system voltage level.

Insulation Fault

An insulation fault on the solar module or the string wiring should blow the fuse or trip the ground fault detector. If you suspect that you have an insulation fault, then an insulation test meter should be used in locating the fault. If the solar module has an insulation fault then it should be replaced, the module can not be repaired.

The typical causes of insulation failures are:

- Animals chewing on insulation.
- Sunlight aging of the cable.
- Embrittlement.
- Cracking.
- Moisture ingress.
- Freeze and thaw cycles.

Reverse Current

Reverse current can flow if there are voltage differences between the parallel connected strings. The fuse protects against this when the reverse current gets too large. Identify the reason why the string is producing a low voltage and repair.

Shading effects and solar modules facing in different directions from others in the system can cause this problem to occur. It can be prevented by using blocking diodes in the individual strings.

Aging

Solar modules will be affected by age. It depends on the module manufacturer and the technology used, but typically a 1% loss in power per year is normal for a solar module. So after twenty years, it will generally only output about 80% of its rated STC power in STC conditions.

On old solar modules, you should not be surprised to see them looking brown. This is simply the EVA encapsulant aging.

Module Aging

Modules can age and the encapsulant slowly changes color. This can be seen in a system that has both old and new solar modules in it.

String Testing

Safety note: Never remove a DC fuse or break a DC connection under load, you may cause a fire.

Design note: It is for string testing purposes that it is not recommended during the design of a solar power system that more than one string be connected to an individual string fuse. On small one or two string systems, string fuses are not required but are desirable for individual string testing purposes.

The solar module strings are generally tested once per year. For this, you will need the following tools:

- Solar module data sheet
- Thermometer
- Irradiance meter
- Clamp on DC ammeter
- DC voltmeter
- Live voltage protective safety equipment
- Insulated fuse puller

The tests that are performed look for the following:

- String open circuit voltage (Voc)
- String maximum power point voltage (Vmpp)

- String maximum power point current (Impp)

Before performing the tests, make sure that the solar modules are clean. These tests are performed at the closest location to the string which will generally be the inverter in a residential system, and the combiner boxes in the larger systems. You will need a clear day with no clouds near to the sun. Stable conditions are important for obtaining accurate results.

String Open Circuit Voltage (Voc)

String open circuit voltage (Voc) testing is performed by turning off the inverter or the DC disconnect for combiner boxes. Verify that the strings are not passing current with the clamp on ammeter. Remove the fuse for the individual strings with the insulated fuse puller. With a DC voltmeter, measure the voltage on each string. Measure the solar module cell temperature. Using the solar module data sheet, adjust for:

- Measured solar cell temperature
- Age

This should match the voltage measured on the strings. If the string voltages are reading differently, then you will have a problem with the low voltage string(s) that will need troubleshooting.

String Maximum Power Point Voltage (Vmpp)

String maximum power point voltage (Vmpp) testing is performed with the system in operation. Measure the voltage of the strings with the DC voltmeter. Measure the solar module operating temperature. Referring to the solar module data sheet, adjust Vmpp for:

- Measured solar cell temperature
- Age

The measured voltage should approximately match the calculated value for Vmpp. If not then you will need to start troubleshooting the string(s).

String Maximum Power Point Current (Impp)

String maximum power point current (Impp) testing is performed with the system in operation. String maximum power point current (Impp) can be measured with a DC clamp-on ammeter with an accurate range of 0 to 20 amps. It is important with a DC clamp-on ammeter that it is consistently applied to the circuits with the current flow passing through it in the same direction. DC current is proportional to irradiance, so measure this. Measure the solar module cell temperature. Referring to the solar module data sheet, adjust Impp for:

- Measured irradiance
- Measured solar cell temperature

- Age

This should be approximately the current that is measured on the string. If not, you will need to troubleshoot the string.

Record all values measured in the maintenance logs for the solar power system.

Solar Module Testing

Generally you will test replacement solar modules prior to their installation. For roof mounted systems and other systems with poor access, you should verify that each module is functional before installation due to the difficulties in troubleshooting these systems. Simply expose each of the solar modules to the Sun and ensure that they are producing the temperature corrected voltage with a multimeter prior to mounting it.

If you have the competence, you could also do a short circuit test by connecting the positive and negative leads together while the module is covered with an opaque cover. Use a DC clamp meter to check the current from it when exposed to the Sun. The temperature corrected short circuit current will be proportional to the irradiance. Cover up the module with an opaque cover to undo the leads prior to installation.

The bypass diodes can be checked by covering up each internal string in the module. You should notice a drop in voltage that corresponds to the number of cells in the internal string. It should still produce the same current in a short circuit test.

Mounting System O & M

There are generally three ways that your solar modules will be mounted:

- Fixed tilt
- Single axis tracker
- Dual axis tracker

The fixed tilt system is the most common and is widespread. The solar modules are either mounted to a roof, building or are ground mounted in a fixed position inclined to face south at a tilt angle matching the latitude. Some systems allow you to adjust the tilt angle of the modules for the season, but it appears that most people prefer the low maintenance option of mounting the modules into a fixed position for the entire year. The fixed tilt system is the most reliable configuration and also the lowest cost. The downside is it has the lowest annual energy output of the mounting systems.

The single axis tracker works well in. The solar modules are mounted on a rotating north-south axis which allows them to track from east to west during the day. There are two types of single axis trackers generally available. The first has the north-south axis mounted horizontal and the modules can track in the east to west direction. This system works well in or near to the tropics where the Sun can be almost directly overhead. The second has the north-south axis inclined to match the latitude and this enables the solar modules to face the Sun in spring and fall. This system works better as you move further away from the tropics. A single axis tracker can increase power output by about 20% when compared to a fixed tilt system. The single axis tracker does not cost much more than a fixed tilt system and the extra expense is generally offset by the extra annual energy yield of the system, particularly in southern USA locations.

The dual axis tracker has the modules tracking the Sun from sunrise to sunset, keeping the solar modules in the optimal position for maximum power generation. A dual axis tracker can increase annual energy output by about 30% when compared to a fixed tilt system. The downside to a dual axis tracker is that it requires a lot of space, can be very tall, has a complicated control system, is expensive, and it is the highest maintenance system.

The diagram on the next page shows the differences for each tracker system at noon with the seasons.

Complete Solar Photovoltaics © Steven Magee

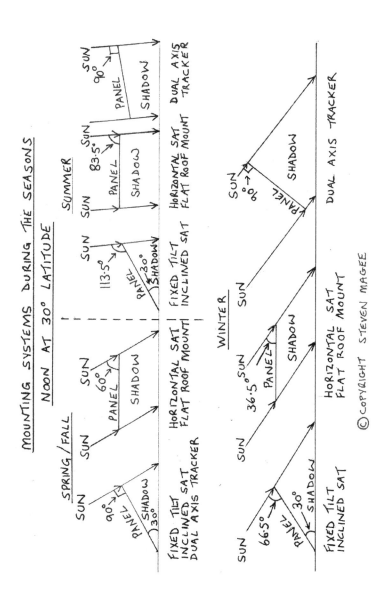

Fixed Mounting O & M

The fixed mounting needs little maintenance and the following is recommended:

- A visual check of the mounting system.
- If it is adjustable for the seasons, then you will be making adjustments to the tilt angle every three months.
- Repair any corrosion on the mounting system with the appropriate technique recommended by the manufacturer.
- Inspect for loose screws, nuts and bolts and tighten as needed.
- Tie up any loose cables as needed.
- Check the fencing around the system if it has this.
- Make sure the grounding system is intact.

Fixed Tilt Servicing

You may need specialized equipment to service a fixed tilt installation, such as this car park canopy.

Single Axis Tracker O & M

The single axis tracker has a control system. In addition to the maintenance detailed for the fixed system you will need to do the following:

- Turn off the control system, open the control system box and tighten all connections.
- Check that the encoding system, usually an inclinometer, is in good condition.
- Check the axis travel limits function correctly.
- Inspect the bearing systems and lubricate as needed.
- Check that there are no parts catching on the moving mechanisms by driving the system through the full range of motion.
- If it has a hydraulic drive system then check the condition of the hoses and look for leaks.
- Replace any consumable parts or fluids on the system according to the manufacturers recommendations.

If the system is pointing in the wrong direction then it will most likely either be:

- A blown fuse.
- A broken encoder.
- A broken drive system.

- It may have the wrong time on its clock.
- It may think that it is in a different global location.

Diagnose the problem and fix accordingly.

Complete Solar Photovoltaics © Steven Magee

A Single Axis Tracker

This single axis tracker was being maintained. You get a sense of the scale of the system when comparing it to the maintenance truck.

Dual Axis O & M

Dual axis systems need all of the maintenance as detailed above. The only significant difference is that there are two axes instead of one.

Tracking System Drives

Tracking system drives are generally either electrical or hydraulic rams and you may need specialized equipment to service them.

Cable O & M

For all outdoor cabling exposed to the Sun, solar photovoltaic DC rated cable should have been installed due to its resistance to ultraviolet radiation. Once inside a building, conduit, or underground this requirement need not apply.

On ground mounted installations it is important to remember that animals will have been able to access the cables and equipment, so you will need be familiar with your local wildlife in the area. Hopefully the installer will have enclosed all accessible cabling with protective covers to prevent this.

Do a thorough inspection of all cables and connectors. Generally it will be insulation damage that you will be looking for. Replace any cable ties as needed.

Look for any signs of overheating on the cabling and connectors. If you are seeing signs of overheating or insulation failures, then you may need to rewire the system to a larger cable and / or connector size. Performing the calculations for the system and de-rating the cables for high ambient temperatures will help with diagnosing an overloading problem.

For locating cable faults, particularly underground, a time domain reflectometer (TDR) can be extremely useful on large installations.

Complete Solar Photovoltaics © Steven Magee

Solar Array Cabling

Most solar array cabling damage occurs where the cabling is accessible to animals.

Combiner Box O & M

The combiner and re-combiner boxes are a low maintenance item. Turn off and lock off the DC disconnect prior to performing any work. Cover the solar modules if you can to prevent them from producing power. On a large system, it will be impracticable to cover all solar modules feeding into the combiner box, as there may be hundreds of them feeding into the box. In this case either take live electrical working precautions or perform the maintenance once the Sun has gone down.

Look for the following in combiner boxes:

- Blown fuses / tripped circuit breakers.
- Cracked or exploded fuses.
- Overheating fuses / circuit breakers.
- Overheating fuse holders.
- Overheating cabling.
- Overheating connectors.
- Loose connections.
- Components discoloring or aging.
- Melting enclosure boxes if made of plastic.

If you see any of the above signs, then this may be an indication that the combiner or re-combiner box does not meet the electrical current, interrupt current and / or voltage requirements of the circuit. Run the electrical calculations for this circuit and de-rate the cables and components for high ambient box temperatures. Check that all components meet the DC voltage, interrupt current, and normal operating current requirements of the circuit.

Assuming that the box is looking to be in good condition then perform the following tasks:

- Tighten all connections
- Check that all fuses are in good condition, none blown
- If the box has blocking diodes installed, then check these with a diode tester to make sure that they are functioning.
- Lubricate any parts as needed
- Repair any signs of corrosion following the manufacturers recommendations
- Any other tasks that the manufacture recommends
- Update the maintenance logs for the system

You normally will find the combiner boxes in amongst the solar array. They are generally under the solar modules and are shaded by them. This is shown in the following picture.

Complete Solar Photovoltaics © Steven Magee

Combiner Boxes

It is normal to find the combiner boxes in the solar arrays.

Inverter O & M

Inverters come in many different types and sizes. This book is dedicated to grid tie inverters and this is what we will consider. All residential and commercial solar photovoltaic grid tie inverters have to comply with UL1741 in the USA. This ensures reliability and performance from them. As such, the inverter should perform for many years without any problems.

The inverter is considered to be a consumable item and it is likely after a ten year life of operations that it will start to wear out, just like an average car will. Providing it with regular maintenance should give it a prolonged life span.

The inverter is a high power item and you will need to exercise caution around it when working on it. There is a lot of protection built into an inverter and it will generally give an alarm code if it is having any problems.

Inverters can suffer from over and under voltage problems in both the DC and AC circuits. Causes of the voltage problems in the DC circuit are:

- Incorrect solar string sizing.
- Fuses and circuit breakers overheating.
- Undersized DC cabling.
- Solar module failures.

- Aging effects.

The AC circuit problems usually come from:

- Excessive volt drop on the AC cabling. You may need to increase your cable size to prevent the inverter from tripping on high AC voltage.
- Large number of solar systems installed and causing problems on the grid. You may trip out the either the inverter or the grid if the amount of solar installed in the area is too high and you will need to work with the utility in order to rectify this problem.
- Poor grid power quality. Inverters will disconnect from the grid when power quality issues arise on it. They will not reconnect until the problems have gone and the grid has stabilized. If your inverter keeps disconnecting from the grid, you will need the utility to check their system for power quality issues.

The items that we will look at on the inverters are:

- Air filters.
- Electrical Terminals.
- Cables.
- Output power.

Before opening the inverter to perform work on it, turn off the AC disconnect and DC disconnect and lock them off. Open the cabinet and perform the following tasks:

- Verify that the DC circuit is off using a volt meter.
- Verify that the AC circuit is off using a volt meter.
- Inspect the interior of the inverter for any evidence of wildlife.
- Dust out the cabinet as needed.
- Inspect for any component damage.
- Change the air filters.
- Tighten all electrical connections on the DC and AC cabling.
- Verify that the grounding system is intact.
- Any other tasks that the manufacture recommends.

Check that all tools have been removed from the cabinet and close it. The AC and DC disconnects can now be closed and the inverter switched back on.

Measure the temperature of the solar modules that are connected to the inverter. With an irradiance meter, take an irradiance reading. Referring to the solar module data sheet, adjust the maximum power point rating for irradiance and temperature effects. The inverter should be producing AC power that is proportional to the adjusted DC power that is connected to it, minus several percent for inverter DC to AC conversion inefficiencies and DC circuit inefficiencies.

If it is not producing the expected power, check the DC circuits that are connected to it.

Finally, update the maintenance logs for the system.

A typical residential inverter is shown on the next page. The following pages show a typical commercial and utility inverter system.

Complete Solar Photovoltaics © Steven Magee

Residential Inverter System

Residential inverters are typically quite compact.

Complete Solar Photovoltaics © Steven Magee

Small Commercial Inverter System

This is a small commercial inverter system with its transformer. The disconnects are behind the inverters.

Complete Solar Photovoltaics © Steven Magee

Large Commercial / Utility Inverter

This inverter is typical of what you will see on the larger installations.

Switchgear O & M

The AC low voltage switchgear is low maintenance with few checks. It is important to make sure that the power is turned off and locked off before performing maintenance. The source of power needs to be disconnected from the grid and also from the inverter system that feeds the switchgear. Remember that in a solar photovoltaic system we have at least two sources of power. In larger systems there may be other sources of power to consider such as emergency generators, uninterruptible power supplies (UPS), battery systems and so on.

After verifying that the system is safe for maintenance the following checks should be performed:

- Check the cable connections are tight.
- An inspection of the cable insulation for damage.
- Look for signs of overloading.
- Repair any corrosion to the enclosures using the recommended techniques in the manufacturers manual.
- Look for any signs of wildlife inside the enclosures.
- Lubricate the mechanisms according to the manufacturers requirements.
- Any other manufacturer recommendations.
- Update the maintenance logs for the system.

Distribution O & M

Distribution operation and maintenance covers the following items

- Transformers
- Distribution poles
- Medium voltage switchgear
- Relays
- Harmonics
- Dirty Electricity

You should be following safety procedures on this equipment as it has the potential to be fatal.

Transformers

Dry type transformers need very little maintenance. Turn off and lock of the equipment. The following checks should be done:

- Check connections are tight
- Inspect cables for overheating effects
- Check cable insulation for damage
- Dust and clean

- Any other recommendations that the manufacturer requires
- Update the maintenance logs for the system

If you have an oil filled transformer then you will need to do the following additional items:

- Take an oil sample for testing if needed
- Top off the oil level if needed
- Check for leaks
- Any other recommendations that the manufacturer requires

When in service, take a temperature reading of the transformer(s) and if you have a thermal camera, take a thermal image of it.

Distribution Poles

Check the condition of any distribution poles that your system uses. The following items need to be checked:

- Termite or rot damage
- Physical damage
- Birds nesting
- Trees and vegetation overgrowth.
- Visual inspection of any pole mounted equipment

- Any other recommendations that the manufacturer requires

Arrange for power pole and line maintenance if needed and update the maintenance logs for the system.

Medium Voltage Switchgear

The medium voltage switchgear needs to be switched off and locked off. Test to make sure that it is not energized before commencing work. The following items need to be done:

- On vacuum circuit breakers check that the vacuum is good
- On gas insulated breakers check that the gas pressure is good
- Check connections are tight
- Inspect cables for overheating effects
- Check cable insulation for damage
- Dust and clean
- Any other recommendations that the manufacturer requires
- Update the maintenance logs for the system

Protective Relays

The following should be done with the protective relays:

- Check relay settings are correct.
- Check connections are tight.
- Inspect cables for overheating effects.
- Check cable insulation for damage.
- Calibrate if necessary.
- Dust and clean.
- Check the batteries, charging system, and electrolyte levels.
- Any other recommendations that the manufacturer requires.
- Update the maintenance logs for the system.

Harmonics

Inverter systems create harmonics. If you have the ability to check the harmonic levels then this is a good thing to do. Normally, total harmonic distortion (THD) should below 3%.

If not, then this would indicate a problem at the AC output of one of the inverters connected to the system. This would then have to be checked at the individual inverter level to determine which inverter was producing the harmonics.

If you have really severe harmonics on the system you will need to start troubleshooting it. Severe harmonics may cause the wiring on the utility system to radiate radio waves!

Dirty Electricity

If you are detecting dirty electricity and it is not originating at your inverter system, then you may find that it is coming from these places:

- Electronic systems
- Gas discharge lighting
- Arcing connection
- The utility grid

Dirty electricity will show up as noise and distortion on the 60 Hz sine wave. You will need to work through the equipment to isolate the source. The easiest way to establish if the dirty electricity is on the electrical utility grid is to disconnect your equipment from it by turning it off at the main disconnect switch. Measure the utility sine wave. If the distortion is still there, then you should report it to the utility.

If the distortion disappears when you disconnect from the electrical utility grid, then you will need to troubleshoot your equipment until you find the item that is generating the dirty electricity. You should turn everything off and then start switching the items on one at a time until you detect it. The last item that you switched on will be the source of it.

Complete Solar Photovoltaics © Steven Magee

Dirty electricity is a serious health problem, as it will make the electrical wiring start to emit wide-band radio waves. This could induce radio wave sickness (RWS) or electromagnetic hypersensitivity (EHS) into anyone who is near the radio fields that it may generate.

A simple circuit is shown on the following pages which can help with preventing dirty electricity. It is simply a selection of different size capacitors that are connected in parallel at the fuse board. The capacitor plates are all different sizes in order to filter different frequencies that may be present on the AC system.

The simple filter should cause any radio frequencies to be filtered out at the fuse board. It should be mounted in a metal enclosure, as it will be radiating radio frequencies if it is filtering them from the utility network.

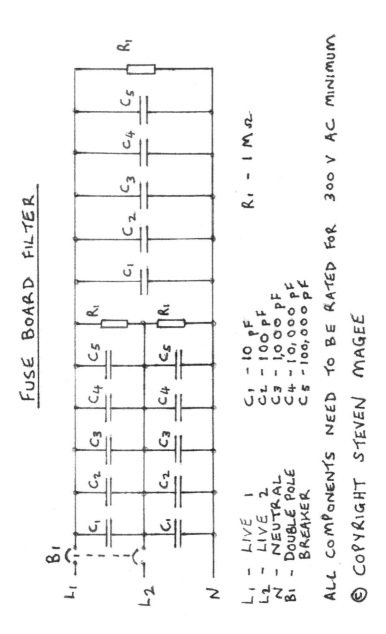

Complete Solar Photovoltaics © Steven Magee

Fuse Board Filter

This is what the filter looks like when it is built.

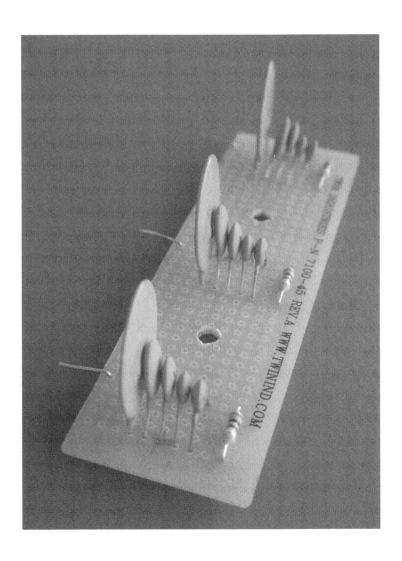

Ancillary Systems O & M

These systems are not essential and are generally installed on larger sites, particularly if they are unmanned:

- Weather Station.
- Security System.
- Video Surveillance.
- Metering.
- Inverter Monitoring.
- String Monitoring.
- Monitoring Services.

Weather Station

Many solar sites install weather systems, particularly if they are unmanned. They can provide useful information that can be used to verify the power output of the system. Most solar site weather systems include the following sensors:

- Reference solar cell with temperature.
- Irradiance sensor.
- Solar module temperature sensor.
- Ambient temperature.
- Humidity.

- Wind speed and direction.
- Rain gauge.

Lets look at each of these in detail:

Reference Solar Cell With Temperature

This reference cell will be made from the same solar cell type as the installed solar modules contain. It is a factory calibrated solar cell. It has a temperature probe attached to it to record the cell temperature. This allows the cell reading to be adjusted to the correct value.

The cell needs regular cleaning to maintain accurate performance and a calibration check.

Irradiance Sensor

This is used to record the irradiance levels and is sometimes referred to as a pyranometer. It is important to keep it clean and to perform a calibration check on it.

Solar Module Temperature Sensor

The solar module temperature sensor is attached to the rear surface of one of the installed solar modules. This allows the operating temperature of the solar module to be known and for the solar module output to be adjusted for temperature effects. A regular calibration check should be performed on this.

Ambient Temperature

The site ambient temperature is recorded for historical purposes. A regular calibration check should be performed on this.

Humidity

The site humidity is recorded for historical purposes. A regular calibration check should be performed on this.

Wind Speed and Direction

The wind speed and direction can be used to assess the cooling effects from this on the solar modules. A regular calibration check should be performed on this.

High wind speeds are a concern for system damage from flying debris. After a high wind speed event has been detected, a site visit should be arranged to check on the equipment.

Rain Gauge

It is useful to know the historical rainfall, so that you can schedule cleaning. If you have a site with a consistent rainfall, then it is unlikely that you will need to clean the solar modules regularly. If you have a site that only gets sustained rainfall on one or two occasions per year, you may want to schedule solar module cleaning more frequently. A regular calibration check should be performed on this.

If any part of the weather system fails, then it is usually replaced. Weather system parts are generally regarded as consumables due to the harsh environment that they work in.

A typical weather system is shown on the next page.

Complete Solar Photovoltaics © Steven Magee

Weather System

A weather system mast is typical on the larger installations.

Security System

On remote sites a security system is usually installed. The security system will usually use beam break sensors within the caged areas of the solar site. Keep these beam break sensors clean to ensure that false alarms do not occur.

Thefts have not been a significant problem in the industry yet. It helps that the solar modules are firmly attached to their mounting structure and that tools are needed to remove them. Generally most thieves are after the copper wiring. This is why it is important to always check the grounding systems of your installations, as they may have been stolen.

As public understanding of solar photovoltaic power systems increases and demand for the solar modules goes mainstream, this may develop. It is a good idea to install the security system during construction of the project to keep your insurance premiums down on the system.

If the security system fails, try cycling the power to it. You may have to disconnect the battery back up for this to be successful. Check the settings if it is still not working. The next step would be to replace the troublesome part of it.

Video Surveillance

On remote sites usually internet based video surveillance cameras are installed with the pan, tilt and zoom (PTZ) feature. These will be low light level cameras that can work under moonlight. They are mounted in weather resistant housings. The maintenance that will be needed is just a regular cleaning of the camera housing.

If the camera fails, then try cycling the power to it. Check the internal settings. If this doesn't work, replace the camera.

Metering

The electrical meter for the solar photovoltaic system can be connected to the internet. This is useful for billing purposes if the energy is being sold to a customer. It also is very useful for detecting a problem at a remote site. If the site is not producing energy, then it can be quickly detected by monitoring the meter. A regular calibration check should be performed on this.

If the meter fails, try a power cycle on it. Check the internal settings on it. If this doesn't fix it, then replacement would be the next step.

Inverter Monitoring

Most inverters have the capability to be monitored and controlled. This is useful for power factor control and for detecting inverter faults. If the inverter has power factor control, then it can be used to provide power factor control to the grid that it is connected to. The utilities like to have this feature if it is available.

The inverters have many fault codes and it is useful for a remote site to know the fault code in order to schedule a repair visit.

A nice feature that the inverters have is the ability to monitor the DC inputs to them. This is commonly referred to as "zone monitoring". When monitoring at the zone level, it is easier to detect problems with the solar module strings by comparing the zones to each other. If one zone is consistently reading less than another equally sized zone, then you will need to schedule a repair visit to that zone.

If the inverter monitoring fails, a power cycle of the inverter will usually fix it. Check the internal settings if is is still not working. If the problem persists, then the monitoring circuit board will likely need to be replaced.

String Monitoring

Monitoring can be done at the string level. On a residential system this is normal for the inverter to be monitoring at this level due to the small number of strings connected to it. On commercial and utility systems, this would be installed in the combiner boxes.

On utility systems with several thousand solar photovoltaic strings, this is a lot of monitoring and as such, it is also expensive to install and maintain. It is nice to have, but not really needed. Good inverter zone monitoring should be able to detect an individual string failure.

If you have string monitoring installed, then it generally will only need attention when when it fails to report string current. Usually, this requires either a reset of the computer module or replacement of the computer module.

Monitoring Services

There are many companies now offering monitoring services for solar power generation systems. While not so useful for residential and commercial applications where there will be people present to check on the system, they are extremely useful for remote sites.

The monitoring center will send out automatic email notifications to you when problems are detected on site. This enables a response to be arranged quickly before too much energy generation is lost. On a large solar photovoltaic power generation system, it doesn't take much time of the system being out of service before thousands of dollars of energy generation has been lost.

These monitoring services are invaluable on large generation systems and their expense is far outweighed by the revenue saved by keeping the solar power generation system in full service.

Summary

Solar photovoltaics is a very reliable technology. The field is set for growth and systems should become commonplace with the support of the government incentives. When designed and installed correctly on a good site, solar photovoltaic power systems will reliably produce energy for many years with minimal maintenance.

In the future it is quite possible that every new home constructed will come with a solar photovoltaic power system as standard. Installing solar photovoltaics during the construction phase of a home is the most cost effective way to adopt the technology.

Many companies have solar photovoltaic roofing products on the market and in development to meet the future needs of modern, self contained buildings. Net zero is probably the future of green construction, where a building will produce its own energy needs with an annual energy use of zero from the utility grid.

With the development and adoption of AC solar module technologies, solar photovoltaics will not need any specialized knowledge to install it and any electrician will be able to work with it. AC solar photovoltaic modules will be just like an appliance that will plug into the house electrical system. Indeed, in the future you may well purchase your solar photovoltaic modules at the appliance shop.

The smart grid is in development to allow these distributed generation systems to be effectively managed by the utilities. The electrical grid will be quite different in the future with power generation taking place on the majority of consumer premises with the utility grid supporting it when sufficient power cannot be generated due to climatic conditions.

The utilities will continue to build large solar photovoltaic power plants and they will keep getting larger with each generation of plant. The utilities are currently learning how to construct solar photovoltaic systems that work well with their existing infrastructure. Today's utility systems are paving the way towards the very reliable solar photovoltaic power generation systems of the future.

Solar photovoltaics is a rapidly changing field and new technology is constantly being developed. As such, always design from the latest electrical and building codes. In the USA, the National Electric Code Section 690 Solar Photovoltaic Systems details the latest electrical design standards. The USA National Electric Code is updated every three years and it is important to design from the latest version.

Follow the design and installation manuals for the products used in your design. Always have any designs independently verified by a registered Professional Engineer that specializes in solar photovoltaics.

Solar photovoltaics is a very reliable technology and should need little maintenance. A regular check of the system components and monitoring of the system performance should be sufficient. As ever, always follow the manufacturers maintenance schedules for your system.

When parts do break down on your system, you should now be able to diagnose the problems. Remember to work safe on your system, as there can be high voltages and large currents present in grid connected solar power systems. Always follow the instructions in the equipment manuals when working on your systems. When troubleshooting a system you should refer to the latest edition of the solar photovoltaic electrical codes.

It is an exciting time in the solar photovoltaic industry and I hope that this book has given you an overview of where the technology is today and where it may be heading in the future. By reading this book you will now be prepared for entering the field of solar photoltaics and will know how to work with the systems.

I recommend that further reading is done in addition to this book. The reference section is a good source for further information, as are my other books on the subject. This book is not a substitute for a formal training course on solar photovoltaics and this would be a great next step in the learning process.

I would recommend that you read the following books to improve your solar photovoltaic knowledge further:

Solar Site Selection for Power Systems.

Solar Photovoltaic DC Calculations for Residential, Commercial, and Utility Systems.

I also have a collection of reference information that you may find useful. It also shows the NREL charts for the USA in color which a solar photovoltaic designer will frequently reference. It is the information that I most frequently use in working with solar power systems compiled into one book:

Solar Photovoltaic Resource for Residential, Commercial, and Utility Systems.

I hope that you have enjoyed this book and welcome to the world of solar photovoltaics!

References

- Home Power Magazine: http://homepower.com/
- NFPA National Electrical Code 2005, 2008, 2011
- IEEE/ANSI-C2-2007 National Electric Safety Code
- National Renewable Energy Laboratories: www.nrel.gov
- Occupational Safety and Health Administration: www.osha.gov
- OSHA Standard 1910.269: Electric Power Generation, Transmission, and Distribution
- United States Department of Energy: www.energy.gov
- Solar Photovoltaic Design for Residential, Commercial, and Utility Systems by Steven Magee
- Solar Photovoltaic Operation, and Maintenance for Residential, Commercial and Utility Systems by Steven Magee
- Solar Photovoltaic Training for Residential, Commercial, and Utility Systems by Steven Magee
- Solar Pro Magazine: http://solarprofessional.com/
- University of Oregon Solar Radiation Monitoring Laboratory: http://solardat.uoregon.edu/SunChartProgram.html
- Wikipedia: www.wikipedia.org

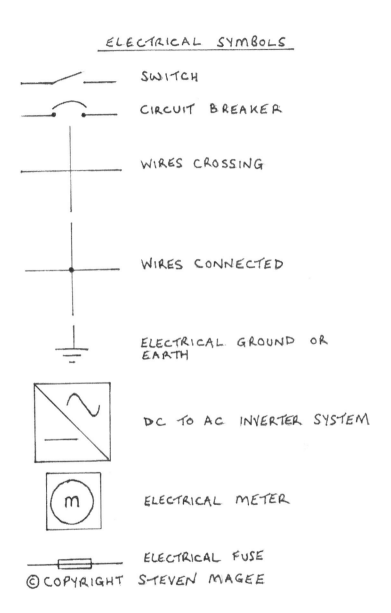

ELECTRICAL SYMBOLS

- ─▭─ RESISTOR
- ─┤├─ CAPACITOR
- ─⌒⌒⌒─ INDUCTOR
- ⋀ CURRENT TRANSFORMER (CT)
- ⋀Y POTENTIAL TRANSFORMER (PT)
- ↑(circle) CURRENT SOURCE
- ⟪⌢⟫ MEDIUM OR HIGH VOLTAGE CIRCUIT BREAKER
- ─−|+─ BATTERY
- ─▶|─ DIODE
- ─−▭+─ SOLAR MODULE
- ─−|+─ (with arrows) SOLAR CELL
- ─⫞⫠─ TRANSFORMER

© COPYRIGHT STEVEN MAGEE

ELECTRICAL SYMBOLS

- CENTER TAPPED TRANSFORMER
- DUAL SECONDARY TRANSFORMER
- DIRECT CURRENT (DC)
- ALTERNATING CURRENT (AC)
- POINT OF INTERCONNECTION
- TRANSFORMER
- POTHEAD

© COPYRIGHT STEVEN MAGEE

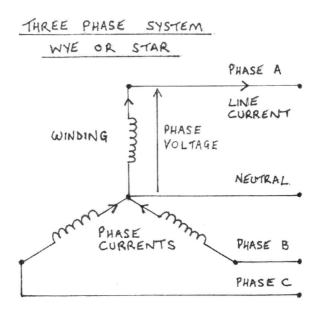

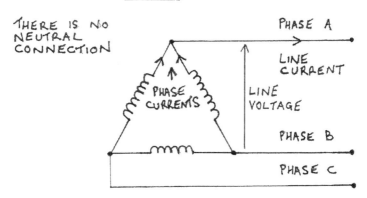

Electrical Formulae

DC minimum fuse sizing = 156% x short circuit current

DC minimum fuse = 156% x I_{sc}

Volt drop = (2 x length x resistance per 1,000 feet x current) / 1,000

Volt drop = (2 x L x R x I) / 1,000

Power = volts x current x cosine phase angle

$P = V \times I \times \cos \phi$

Note: Three phase calculations are all for a balanced circuit.

Three phase power = square root of 3 x line volts x line current x cosine phase angle

$P = \sqrt{3} \times V_L \times I_L \times \cos \phi$

Three phase line current = phase current x square root of 3

$I_L = I_P \times \text{sqrt } 3$

About the Author

Steven started his career at one of the largest university research hospitals in Europe. Working in the electrical engineering group, he obtained a Bachelors with Honors in Electrical and Electronic Engineering. Human health was a strong draw and he moved into the biomedical team, serving the regions hospitals. During this time he developed a fascination for human disease and the causes of it, many of which were not understood.

He joined the Isaac Newton Group of Telescopes in 1999 and went to live in La Palma. La Palma is part of the Canary Islands, governed by Spain. During this time he worked with the leading European astronomers and developed his astronomical and optics skills. He became fluent in Spanish and their culture.

In 2001 he became a Chartered Electrical Engineer and joined the W. M. Keck Observatory in Hawaii. This was the world's leading astronomical facility and home to the world's two largest segmented mirror telescopes. Steven developed segmented optics and interferometry skills while working alongside world leading astronomers. During this time Steven constructed his own off-grid solar powered home in the last of the traditional Hawaiian fishing villages in Miloli'i, Hawaii. He learned Hawaiian Pidgin English and the Hawaiian culture during his time there.

Complete Solar Photovoltaics © Steven Magee

In 2006, Steven became the Director of the MDM Observatory in Sells, Arizona, USA. Working for Columbia University and later, Dartmouth College, he developed the facility to modern standards. He learned an appreciation of the native Americans and their culture from the Tohono O'odham Nation.

In 2008, Steven joined the solar power revolution that was sweeping the USA and commissioned the largest CIGS thin film solar photovoltaic installation in the world.

A year later he commissioned the largest solar photovoltaic power plant in the USA. The system rated power was quoted as 25MW with over 90,500 solar modules that were mounted to 158 single-axis tracker systems in three hundred acres of land.

He went on to develop the solar photovoltaic team for a large international company.

In 2010 he started to research radiation and publish the leading books on the subject. Some of the discoveries that Steven made during his independent research are:

Human Health:

- Found the cause of cancer (incorrect human environmental conditions).
- Found the preventative measures for cancer (correct human environmental conditions).

- Found links between solar radiation and human health issues.
- Found links between artificial radiation and human health issues.
- Developed the cause of allergies to pollen (Pollen Deficiency).
- Developed the causes of allergies to sunlight (Sunlight Deficiency, filtering, interference, and overexposure).
- Developed the causes of allergies to nature (Nature Deficiency).
- Found that the human has the genetics of an outdoor forest animal.
- Found that the human mind and body does not perform correctly away from the presence of a tree canopy.
- Discovered extensive photosynthesis effects are taking place within the human body.
- Characterized the different human skin types and their solar radiation environment.
- Electrical poisoning from anti static devices (ASD) and grounding (earthing) systems.
- Developed the field of plant growth defects in the areas of electromagnetic interference, AC stray voltage and currents, solar radiation, toxic light, and nighttime interference radiation.
- Linked the cause of sick building syndrome (SBS) to coated double glazed glass (Low-E), artificial light, de-mineralized water, contaminated air, contaminated food, electromagnetic interference (EMI), and AC stray voltage effects.

- Developed the field of human aggression (environmental pollution).
- Interpreted the famous 1991 Barbury Castle crop circle: Energy systems are causing interference and the wheel is in motion for cellular growth to turn off.
- Widespread radiation sickness in modern human society from the effects of unnatural solar radiation, man-made radiation, and electromagnetic interference.
- The radiation detoxification (hibernation).
- Discovered that electromagnetic interference is a stimulant to the human mind and body, just like alcohol and drugs.
- Characterized electromagnetic interference exposure in the human body: Headaches, increased sexual desire, changed personality, aggression, and fatigue. Over exposure leads to general illness, arthritis, depression, dementia, mental illness, and cancer. These are documented as Radio Wave Sickness (RWS) and Electromagnetic Hypersensitivity (EHS) in the medical community.
- Found that the human mating and fertility cycle is triggered by seasonal electrical storms.
- Found that the modern human is overloaded on electromagnetic interference from daily exposure to it and is unnaturally in a constant state of sexual desire.

Complete Solar Photovoltaics © Steven Magee

Electromagnetic Interference:

- Found that the electrical distribution system is in the process of turning into an unlicensed wide-band radio transmitter due to a high penetration of electronic products. This may cause radio wave sickness and electromagnetic hypersensitivity to occur in many people.
- Found that many electrical utility poles are emitting large amounts of radio waves into the human environment. They are unlicensed wide-band radio transmitters.
- Excessive electromagnetic interference from many laptop computers.
- Excessive electromagnetic interference from many flat screen digital TV's.
- Electromagnetic interference from electronic power conversion systems (inverters) and the effects on human health.
- Solar photovoltaic (PV) electromagnetic interference and the effects on human health.
- Street light electromagnetic interference premature death clusters.
- Linked electrical fuse board electromagnetic interference to human illness and disease.
- Developed the field of power-line illness and disease.
- Developed the field of human AC "stray voltage" illness and disease.

Complete Solar Photovoltaics © Steven Magee

Solar Radiation and Light:

- Developed the field of "toxic light".
- "Archimedes Death Ray" effect in architecture.
- Unnaturally high levels of solar radiation in modern human society.
- Tree canopy light interference (interference green light).
- Water light interference (reflection and lensing).
- Light modification by plants.
- Radiation suppression by nature.
- Nighttime interference radiation.
- Atmospheric energy interference for solar radiation.
- Satellite, airplane, and structure solar radiation interference.
- Developed the LAMB theory = Light of Alien Moons is Baneful (Alien Moons = man made satellites, airplanes, chemtrails, smoke, pollution, artificial lighting, and so on).
- Developed the LION theory = Light Interference Obliterates Nature.
- The extinction wavelength of light.
- Developed the solar photovoltaic power equations.
- Found the sources of solar photovoltaic power system overloading.

Steven Magee is now one of the leading radiation and human health experts in the world and is providing consulting services to the industry. He is trained in human health, biomedical systems, astrophysics, optics and engineering.

> **"The task of modern educator is not to cut down jungles, but to irrigate deserts."**
>
> C S Lewis.

Author Contact

Steven Magee,
3618 S. Desert Lantern Road,
Tucson,
AZ 85735,
USA

I hope that you found the book informative and please let me know about any questions or comments about the book.

I am a consultant on the areas that I research and please feel free to contact me for any help or assistance.

You may find my other books useful:

Solar Photovoltaic

- **Complete Solar Photovoltaics for Residential, Commercial, and Utility Systems:** Steven Magee has combined his three top selling books on solar power systems into one edition. Complete Solar Photovoltaics will train you on solar photovoltaics and show you how to design grid connected solar photovoltaic power systems. Operations and maintenance is detailed to enable you to have a complete understanding of solar photovoltaics from start to finish.

- **Solar Photovoltaics for Consumers, Utilities, and Investors:** This book details solar photovoltaic systems for consumers, utilities and investors. This would encompass residential, commercial and utility systems that are connected to the utility grid. There is a discussion of the different technologies available for the consumer and their advantages and disadvantages. For the utilities, there is invaluable advice on planning and constructing large projects. For the investor, forward looking statements try to predict the future of solar photovoltaics.

- **Solar Photovoltaic Training for Residential, Commercial, and Utility Systems:** This book details solar photovoltaic training for those who are interested in this area and also for those who are already working in the field. This would encompass residential, commercial, and utility systems that are connected to the utility grid. It is a comprehensive overview of a rapidly growing world of solar photovoltaic power generation technology.

- **Solar Photovoltaic Design for Residential, Commercial, and Utility Systems:** This book details how to design reliable solar photovoltaic power generation systems from a residential system, progressing to a commercial system, and finishing at the largest utility power generation systems. By following the guidelines in this book and your local solar photovoltaic electrical codes, you will be able to design trouble free solar power systems that give many years of reliable operation. When designed well, solar photovoltaic power generation is an excellent source of electrical power that results in much lower electricity bills, the power company will even refund you for the excess energy generated by your system if it is large enough. Building a grid tied solar power system is a relatively easy task. Given the large amount of government and electrical utility financial incentives that are available, it is a great time to join in the solar power revolution that is taking place in the world today.

- **Solar Photovoltaic Operation and Maintenance for Residential, Commercial, and Utility Systems:** This book details how to operate and maintain residential, commercial, and utility solar photovoltaic systems that are connected to the utility grid. By following the guidelines in this book you will be able to operate and maintain solar power systems that should give many years of reliable operation. Invaluable trouble shooting advice will aid in returning your system to full operation in the event of a problem.

- **Solar Photovoltaic DC Calculations for Residential, Commercial, and Utility Systems:** This book details how to run calculations for the DC circuit of solar photovoltaic systems. This would encompass residential, commercial, and utility systems that are connected to the utility grid. It covers the range of conditions that solar photovoltaic modules are exposed to throughout the year and shows how to incorporate these into an effective DC circuit that is well designed and reliable.

- **Solar Photovoltaic Resource for Residential, Commercial, and Utility Systems:** This book is a resource of information that is used in the solar photovoltaic field. This would encompass residential, commercial and utility systems that are connected to the utility grid. It is a comprehensive collection of notes, diagrams, pictures and charts for a rapidly growing world of solar photovoltaic power generation technology. This book is illustrated in color.

Solar

- **Solar Irradiance and Insolation for Power Systems:** This book is a resource of information that is used in the solar power generation field. This would encompass residential, commercial and utility systems that are connected to the utility grid. It is a comprehensive collection of notes, diagrams, pictures and charts for a rapidly growing world of solar photovoltaic power generation technology. This book is illustrated in color.
- **Solar Site Selection for Power Systems:** This book is a comprehensive collection of images, diagrams and notes that document the effects of light and heat in the solar power generation field. This would encompass residential, commercial and utility systems that are connected to the utility grid. This is essential information for a rapidly growing world of solar power generation technology. This book is illustrated in color.

Architecture

- **Solar Reflections for Architects, Engineers, and Human Health:** This book is a comprehensive collection of images, diagrams and notes that document the effects of sunlight in architecture. This is essential information for architects, engineers and the medical profession. The discovery of the "Archimedes Death Ray" effect in architecture is detailed and this book is illustrated in color.

Human Health

- **Solar Radiation – A Cause of Illness and Cancer?** Illness and cancers have become part of our modern culture. It has been discovered that extremely high levels of man-made solar radiation exist in modern society. Could this be the one of the causes of illness and cancers? This book examines the increase in solar radiation and applies it to human health.

- **Solar Radiation, Global Warming, and Human Disease:** This book examines the modern development of the Earth and the potential impacts of this on global warming and human diseases. The destruction of the forests for modern agricultural use appears to have effects that are not fully understood and these are explored. Human health is investigated and the possibility of solar radiation poisoning is evaluated as a causative factor in many diseases.

- **Toxic Light:** Toxic Light takes a look at the light pollution that may be in your local environment and relates it to the health problems that it may cause. Light in the human environment is only just starting to be understood and something as innocent as your sunglasses may be able to make you ill! There are many examples of commonplace items in your environment that may have the ability to affect your health. Get ready for enlightenment about the most important human nutrient of light!

- **Toxic Health:** Toxic Health takes a look at the pollution that may be in your local environment and relates it to the health problems that it can cause. Pollution in the human environment is only just starting to be understood and something as innocent as light may be able to make you really ill! There are many examples of commonplace items in your environment that may have the ability to affect your health. In particular, we will investigate if electromagnetic interference (EMI) is the most toxic thing of all to the modern human!

Religion

- **Solar Radiation, the Book of Revelations, and the Era of Light – Part 1:** Welcome to the Era of Light! Light has long been known to be essential nourishment for the human body. We will explore the different types of light that are present on Earth and relate it to human health and nature. Light is discussed extensively in the Bible and we will see if we can associate our findings to it. Finally, we will investigate if the Industrial

Complete Solar Photovoltaics © Steven Magee

Revolution has created the ultimate toxin of poisonous sunlight!

You can search "Steven Magee Books" for the very latest publications.

www.youtube.com videos supporting the ideas in the books can be found by searching: StevenMageeBooks

"Life-transforming ideas have always come to me through books."

Bell Hooks.

Made in the USA
San Bernardino, CA
22 February 2013